Optimality Theory
An Overview

Explaining Linguistics

D. Terence Langendoen, *Series Editor*

This is the first in a series of volumes of essays designed to introduce and explain major research areas in linguistic theory and practice. It is intended for researchers, scholars and students who have a need or a desire to learn more about these areas. Future volumes in the series will be incorporated as special issues of the journal *Linguistics Abstracts*.

Optimality Theory

An Overview

edited by Diana Archangeli
and D. Terence Langendoen

Copyright © Blackwell Publishers, 1997

First published 1997
Transferred to Digital print 2003

2 4 6 8 10 9 7 5 3 1

Blackwell Publishers Inc.
350 Main Street
Malden, Massachusetts 02148
USA

Blackwell Publishers Ltd
108 Cowley Road
Oxford OX4 1JF
UK

Library of Congress Cataloging-in-Publication Data has been applied for.

ISBN 0-631-20225-0 (hbk.)
ISBN 0-631-20226-9 (pbk)

British Library Catloguing in Publication Data

A CIP catalogue record for this book is available from the British Library.

Camera-ready copy supplied by the editors.

Printed and bound in Great Britain by
Marston Lindsay Ross International,
Oxfordshire

Table of Contents

Notes on Contributors

Diana Archangeli (MA, University of Texas; PhD, MIT) is Professor of Linguistics at the University of Arizona. She taught previously at the University of Illinois, Champaign–Urbana. She is the coauthor with Douglas Pulleyblank of *Grounded Phonology* and is now engaged in a number of projects involving Optimality Theory.

Michael Hammond (PhD, UCLA) is Associate Professor of Linguistics at the University of Arizona. He has also taught at the University of Minnesota and at the University of Wisconsin, Milwaukee. He has published extensively on phonological theory and metrical analysis.

D. Terence Langendoen (PhD, MIT) is Professor and Head of the Department of Linguistics at the University of Arizona. He taught previously at Ohio State University, Brooklyn College, and the Graduate Center of the City University of New York. He has published on a wide variety of linguistics topics.

David Pesetsky (PhD, MIT) is Professor of Linguistics at the MIT. He taught previously at the University of Massachusetts at Amherst, and is well known for his wide–ranging research in syntax.

Douglas Pulleyblank (PhD, MIT) is Associate Professor of Linguistics at the University of British Columbia. He taught previously at the University of Southern California and the University of Ottawa. He is known particularly for research on the phonology of African languages.

Kevin Russell (PhD, University of Southern California) is Assistant Professor of Linguistics at the University of Manitoba. He is currently working on the phonology and morphology of Arabic and Cree.

Margaret Speas (MA, University of Arizona; PhD, MIT) is Associate Professor of Linguistics at the University of Massachusetts, Amherst. She has written extensively on problems of phrase–structure analysis and the analysis of pronominals.

Foreword

The goal in creating this volume has been to offer an accessible introduction to Optimality Theory, a powerful new model of grammar. Our intended audience is anyone with a serious interest in language who desires to understand this model, regardless of their background in formal linguistic theory itself.

What is a grammar and how does it work?

People who know a language are able to produce and recognize a huge number of intricately structured expressions (words, phrases, sentences, etc.). Moreover, they are able to distinguish those expressions which belong to a particular language from possibly very similar expressions which do not. Linguists, the scientists who study language, have assumed that these abilities are accounted for by a mechanism, called a **grammar**, which relates the expressions of a language to the elementary parts of which they are made.

Linguists are thus faced with two related problems. One is to ensure that the grammar of a particular language is able to encompass all of the expressions that can reasonably be supposed to belong to that language. The other is to ensure that the grammar is able to distinguish those expressions which belong to the language from those which do not.

The problem can be compared to that of a fisherman trying to catch in a net all the fish of certain types in a certain area, but nothing else (no other types of fish, no other creatures, etc.). The ideal net would be large and fine enough to gather all the desired fish (the desirables), and be designed to allow the undesired fish and other creatures (the undesirables) to escape. But it may not be possible to construct such a net. Any net which is large and fine enough to catch all the desirables may of necessity also catch some undesirables.

If that is the case, one would need a device (a separator) to remove the undesirables once the catch has been taken, no matter how effective the net is in allowing the undesirables to escape. One might therefore decide to put one's energies more into designing an effective separator than into refining the capabilities of the net to allow the undesirables to escape. The ideal separator is one which always succeeds in removing the undesirables, no matter how many the net retains. If one could design an ideal separator, then one might be content with a net which catches everything in the area, allowing nothing to escape, leaving the job of removing the undesirables entirely to the separator.

The ideal net corresponds to the original idea of a generative grammar (as in Chomsky 1957) that accounts directly for (i.e. generates) *all and only all* the expressions of a given language with no auxiliary devices to remove ungrammatical expressions. Because of the enormous complexity of the grammar which results from trying to put that idea into practice, many linguists chose to drop the *only all* proviso for the generative mechanism (the technical description of this state of affairs is that the grammar **overgenerates**), and to add devices, called **filters**, to eliminate the ungrammatical expressions that the generative mechanism allows; see Chomsky and Lasnik (1977) for a proposal along these lines.

The resulting theory divides the task of separating the grammatical from the ungrammatical sentences to two parts of the grammar: the generative component, which accounts for all the grammatical expressions, allows some ungrammatical expressions, and rejects others (i.e. the net); and the filtering component, which removes all the ungrammatical expressions let in by the generative component (i.e. the separator).

This situation, in turn, has been viewed as unsatisfactory: why have two components of the grammar responsible for separating out the expressions which are ungrammatical in a particular language? Chomsky (1995b:223) states this view as follows:

> The worst possible case is that devices of both types are required: both computational [generative] processes that map symbolic representations to others and output conditions [filters].

In phonological research in the late 1980s and early 1990s, analyses including both generative processes and filters were prevalent. Moreover, in many cases, the same facts might be covered by process or by filter, with no empirical consequences. Optimality Theory was introduced in response to this situation. Optimality Theory opts for the 'ideal separator': a very simple generative mechanism (GEN; see Chapter 1) that allows ungrammatical expressions to be created essentially without restriction, leaving all the work of separating out the ungrammatical ones to filtering devices (EVAL; also see Chapter 1). Because the need was so apparent in phonology, the Optimality Theoretic model has rapidly gained the attention of phonologists worldwide.

In syntactic research, Optimality Theory again is the ideal separator. But the research climate is less receptive to such a model: in general, syntactic analyses have not made rampant use of both processes and filters. For example, Chomsky's Minimalist Program (see Chapter 6) represents a return to the idea of the 'ideal net': a generative mechanism that allows the ungrammatical expressions to escape, permitting only the grammatical ones to be accounted for. Consequently, the Minimalist Program and Optimality Theory can be seen as attempts to avoid the worst-case scenario in opposite ways.

An overview of the book

This book is organized to present an introduction to Optimality Theory, and to demonstrate its workings in phonology, morphology, and syntax. Chapter 1, by Diana Archangeli, first summarizes the goals of formal linguistic research, then introduces

Optimality Theory, showing how it addresses these goals. The reader who has little or no understanding of Optimality Theory would do well to start with this chapter. It serves as a preface to the remaining chapters, since the concepts it introduces are assumed in each of the other chapters. The reader who is already familiar with the basics of Optimality Theory might prefer to go directly to one of the following specialized chapters: Chapters 2 and 3 on phonology, Chapter 4 on morphological issues, and Chapters 5 and 6 on syntax. The book concludes with an Afterword, concerning the nature of the input. A summary of Chapters 2 through 6 follows, including comments on which chapters serve as background for subsequent chapters.

Chapter 2, by Michael Hammond, provides an introduction to syllables and feet, the two central constituents in discussions of prosody. The chapter illustrates how Optimality Theory accounts for a variety of prosodic phenomena. It also provides an excursus into psycholinguistics, with a discussion of how some surprising patterns of speech perception are explained under Optimality Theory, patterns which constitute a serious challenge to derivational models of language. This chapter is particularly useful for the non–phonologist because virtually all of the examples are from English. This chapter relies heavily on the analysis provided in Chapter 1, as well as making use of the theoretical points introduced there; it is also useful to the understanding of Chapter 4, which is about morphology.

Chapter 3, Douglas Pulleyblank's chapter on phonological features, explains the concept of phonological features and illustrates a variety of feature patterns found in different languages. The cross-linguistic sketch of how different languages resolve nasal-obstruent sequences (e.g. *nt*, *ms*, *nb*, etc.) illuminates one of the main advantages of Optimality Theory, its ability to precisely characterize formal differences between languages. The chapter also addresses the issue of how a "segmental inventory" is expressed within a model which allows for no restrictions on the inventory of segments in underlying representation.

Kevin Russell introduces key questions in the study of word formation, or morphology, in Chapter 4. This chapter focuses primarily on the phonological, or pronunciation, aspects of morphology: it does not address the syntactic and semantic reasons why certain morphemes may combine with each other while others may not. Chapters 1 and 2 form a useful introduction to Section 3 in particular, which explores reduplication and infixation phenomena. In Section 4, he turns to English, providing an account of the "multiple use" of the suffix *s* in English. In English, both the possessive and the plural forms of most nouns sound alike: *book's/books*, *tool's/tools*, *judge's/judges*. Interestingly, a possessive plural is formed exactly the same way: *books'*, *tools'*, *judges'*, not *books's* ([bookss] or [booksəs]), etc.

The remaining two chapters explore syntactic problems in terms of Optimality Theory. David Pesetsky begins Chapter 5 with a beautifully clear introduction to the essence of current syntactic theory, elucidating both the phenomena and the formal explanations. This part of the chapter is an excellent introduction for Chapter 6 as well as for the rest of Chapter 5, while the reader who is already familiar with current syntactic theory might wish to skip the introductory section and begin directly with Section 2, comparing standard theory and Optimality Theory in syntax, or Section 3, an exploration of the distribution of *that* and of relative pronouns. In English, we can say *the man **who** I saw*, *the man **that** I saw*, and even *the man I saw*, but we don't say *the man **who that** I saw*.

Pesetsky shows that the facts for the comparable sentences in French are subtly and interestingly different, and provides an Optimality Theoretic account of each pattern.

In Chapter 6, Margaret Speas first evaluates the standard "principles and parameters" theory of syntax, and shows that the inviolable principles of this theory are inviolable simply because each such principle includes an "escape hatch" for when it does not hold. She then shows that by adopting Optimality Theory, the principles can be expressed more generally, the escape hatches being eliminated in favor of constraint ranking. The discussion centers on the analysis of "null pronouns", occurring in the position of the underscore in sentences like *Mary expects __ to promote Bill* and *__ To behave in public would enhance Bill's reputation*. In the first, it can only be Mary who will do the promoting, whereas in the second, the one being admonished to behave might be Bill, but might also be some other person. After formulating constraints and constraint rankings to explain these facts, she analyzes the properties of null pronouns in a number of other languages to show that OT also insightfully accounts for the cross-linguistic patterns these pronouns display.

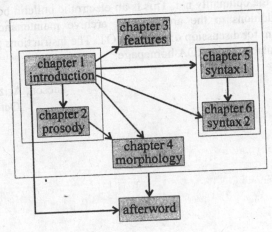

Kudos

A number of people worked very hard to bring this volume to publication. Several of these individuals are identified in the chapters for their contributions to the development of their respective content. We were ably assisted in the formatting, editing, and indexing of this book by Laura Moll–Collopy, Keiichiro Suzuki, and Dirk Elzinga. We are very proud of the results they achieved.

Funding for this book came in part from National Science Foundation grant BNS–9023323 to Diana Archangeli, for which we are grateful.

Finally, for their patience and moral support throughout the duration of this project we thank Dante Archangeli and Nancy Kelly. Special thanks go to Marina and Amico Archangeli for always being there.

Electronic Access to Optimality Theory

Readers who are interested in accessing more material on Optimality Theory have three options. The first is to look to the published literature. A good start is this book. However, at the time this book is going to press, there is very little published work available on Optimality Theory. By contrast, electronic access to a wide variety of works is possible. The Rutgers Optimality Archive (ROA) is a well–maintained electronic repository of unpublished works in OT, which is accessible through the World Wide Web. The ROA includes abstracts of most entries.

> **URL of the Rutgers Optimality Archive on the WEB**
> http://ruccs.rutgers.edu/roa.html

The third option is to join the optimality net. This is an electronic bulletin board which posts information about additions to the archive and archive maintenance. It also occasionally serves as a forum for discussion of issues in OT. The instructions for joining this discussion group are available in the ROA homepage.

Tucson, Arizona, USA
October 1996

1

Optimality Theory:
An Introduction to Linguistics
in the 1990s[*]

Diana Archangeli

Optimality Theory (henceforth "OT") is THE Linguistic Theory of the 1990s. It made its public debut at the University of Arizona Phonology Conference in Tucson in April 1991, when Alan Prince and Paul Smolensky presented a paper entitled simply 'Optimality'. In the spring of 1993, linguists around the world found in their mailboxes a pair of hefty and convincing manuscripts: *Optimality Theory: Constraint Interaction in Generative Grammar* by Alan Prince and Paul Smolensky and *Prosodic Morphology I: Constraint Interaction and Satisfaction* by John McCarthy and Alan Prince. Research in Optimality Theory, especially in the area of linguistics known as phonology (see Section 2), has grown tremendously ever since, and is coming to dominate the world of linguistic research as presented at conferences, workshops, seminars, and colloquia; and the Rutgers Optimality Archive is perhaps the most active and extensive of the various electronic publication outlets in linguistics (see Foreword). The impact of OT in the areas of linguistics outside of phonology has not been as dramatic, but it has been significant, and is likely to rival its impact in phonology before long.

Since OT is a theory of generative linguistics and has had its greatest impact so far in phonology, the next two sections present brief summaries of the goals of generative linguistic theory, and more specifically of the goals of phonological theory. Readers who are familiar with this material can skip directly to Section 3, where discussion of OT begins.

[*]Special thanks to Michael Hammond, D. Terence Langendoen, Dirk Elzinga, Keiichiro Suzuki, and Margaret Speas for their careful reading and suggestions which led to improvements in this chapter. Work on this chapter was supported in part by NSF grant BNS–9023323 to the author.

1 What is Linguistics?

There are two central research objectives in linguistics. The first is to determine and characterize **universal properties of language**, the properties that are shared by all languages. Although the manifestation of a specific universal in a particular language may not be the same as it is in the language next door, such universals are thought to be present in some regard in every language. This leads to the second research objective in linguistics, to determine and characterize the range of **possible language variation**.

> **Linguistics is the study of...**
> 1. **language universals**: the range and type of properties shared in some way by all languages.
> 2. **language variation**: the range and type of variation possible between languages.

By the definitions of language–universal and language–specific properties given above, one might imagine that there is a continuum between the two. The term **markedness** is used to refer to this continuum, with completely unmarked properties being those found in virtually all languages and extremely marked properties being found quite rarely. Language universals must be formulated in a way that is able to characterize this distribution.

The central hypothesis driving generative linguistic research today, due to Noam Chomsky (Chomsky 1965, 1975, 1986; see also Pinker 1994), is that these universals are part of the genetic inheritance of every normal human being. Thus, not only do human beings have an innate ability to learn language, but this innate ability is limited, so that not all strings of sounds can be learned as a language, just as not all strings of words can be put together as a sentence of a language. Universal properties of the world's languages result from inflexibility in this innate language capacity; language variation arises from its flexibility. Linguists use the term **universal grammar** to refer to the innate language knowledge that humans have, including both the flexibility and the inflexibility. In our discussion of Optimality Theory, we will see how the model encompasses both universal and language–specific properties, and how markedness is expressed.

> **Universal Grammar...**
> is the innate knowledge of language that is shared by normal humans — it characterizes both the universal properties of language and the variation tolerated among specific languages

In studying a language, the linguist finds evidence to show that there is a pattern to study, then figures out what the nature of the pattern is, and, finally, determines a formal characterization of the pattern. In each of these efforts, linguists maintain a fairly broad approach. When finding a pattern, the concern is not simply "does this pattern exist?" but also "how does this pattern interact with other patterns in the language?" and "how does this pattern compare to similar patterns in other languages?"

For example, in English there are adjectives like *active, tangible*, and *possible*. A negative form of each adjective can be created by adding a prefix, resulting in *inactive, intangible*, and *impossible*. The linguist notes that the negative prefix takes the form *im-*; which ends with a labial nasal (m), whenever it precedes an adjective which begins with a labial stop (*p, b*), otherwise it takes the form *in-*: *imbalance, impolite*, but *inoperative, intangible, infallible, inviolable*. The prefix, then, is analyzed as having an input form, *in-*, which relates to two different output forms, *in-* and *im-*, depending on the context in which the prefix is placed. (See also Chapter 3 for more about this sort of sound change and Chapter 4 on the standard generative phonology relation between a single input and a variety of output forms.) In characterizing patterns, whether phonological, morphological, or syntactic, linguists try to determine the input form, the output form(s), and the nature of the relation between input and output. Optimality Theory offers a specific view of the nature of that relation.

Studies that focus on a single language explore the patterns that exist within that language. Studies that focus on comparable phenomena across languages examine the range of variation possible within natural human language. By understanding the variation that does occur, we are also able to determine those areas where there is no variation. The more common properties or patterns are thought to be universal, part of our innate language endowment. Not all universals are manifested in the same way in all languages however, due to variation. The more robust a universal is in a particular language, the less marked the language is in that respect. A highly marked property is one which has minimal (or no) claims to universality.

Linguists look for...	to determine...
a. **patterns**	their existence and characteristics
b. **variation**	differences among the patterns of different languages
c. **universals**	the properties that are part of our innate language endowment
d. **markedness**	the robustness of a given property within a language

These methods and goals can be more concretely understood by working through particular language data. For example, consider the phonological universal that words start with a consonant–vowel ("CV") sequence. (Ultimately, we refine this notion in terms of syllables and onsets; for the moment "words start with a CV sequences" is adequate.) In English *sing, like, wish* all start with a "CV" sequence. Languages share this property to different degrees. For instance, in Yawelmani (a language we examine in some detail below) *every* word starts with such a sequence. By contrast, the English pattern shows variation in two ways: on one hand, it allows words to start with more than one consonant, e.g. *stripe, gleam, smooth*, while on the other hand, some words start with a vowel (and no consonant): *apple, important, up*. In this regard, the syllables of Yawelmani are less marked than are those of English.

Within linguistics there are four major subdisciplines: phonology, morphology, syntax, and semantics, defined in (1.1). The first three are topics of chapters in this book. There are other subdisciplines as well, including psycholinguistics, sociolinguistics, and phonetics. However, the four areas mentioned here are the core disciplines within formal linguistics.

Each chapter discusses the application of Optimality Theory in a specific subdiscipline in linguistics; in each, we explore the way in which OT characterizes the universals, variation, and markedness in that subdiscipline. OT began its life as a theory of phonology; this introductory chapter follows suit to a large extent. However, the points made extend to other subdisciplines, as is demonstrated in Chapter 4 for morphology and Chapters 5 and 6 for syntax. In this chapter, sound patterns are used simply as a vehicle for better understanding how the model works.

(1.1)   **The four major subdisciplines in linguistics**
  a.   **phonology**   The study of how sounds combine to make morphemes and words, e.g. *in–active*, but *im–polite*, not *in–polite*
  b.   **morphology**   The study of how morphemes combine to make words, e.g. *act–ing, in–act–ive*, but not *in–act–ing* 'not acting'
  c.   **syntax**   The study of how words combine to make sentences, e.g. *I saw the dog* is good English, *I saw dog the* is not.
  d.   **semantics**   The study of how meanings of subparts combine to make meaning of the whole.

2 An Extended Example: Syllable Structure

To make our discussion of patterns, variation, universals, and markedness concrete, some properties of the cross–linguistic distribution of consonants and vowels of words are illustrated in (1.2), with examples from Hawaiian, English, Berber, and Yawelmani.

A simple pattern of consonants and vowels is found in Hawaiian (1.2a). Hawaiian allows no more than one consonant in a row so we find words like *kanaka* 'man' with three singleton consonants: *kanaka*. However, Hawaiian has no sequences of consonants. In fact, when borrowing words from another language, any consonant sequences are altered to fit the Hawaiian pattern: English *flour* becomes *palaoa*; English *velvet* becomes *weleweka*, etc.

English illustrates the opposite extreme, for it allows long strings of consonants in the middle and at the edges of words, as in *construct* and *sprig*, illustrated in (1.2b). An even more extreme case is illustrated by Berber, a language spoken in Morocco, which does not require vowels at all in its words, *txdmt* 'gather wood' along side *ildi* 'pull'.

Finally, a middle ground is struck in Yawelmani, a Native American language that was once spoken in California (Newman 1944). This language allows at most two consonants in a sequence within a word, as in *xa[th]in*, where the sequence *th* represents two consonants, *t* and *h*. Additionally, Yawelmani tolerates at most one consonant at the beginning and one at the end of a word: *xathin* starts with a single *x* and ends with a single *n*.[1]

---

[1]Since phonology studies the sounds of words, it is important not to get confused by the orthographic conventions of a particular language. For example, the symbol [θ] is used to represent the sound spelled *th* in an English word like *sixth* or *ether*. The symbol sequence [th] as in *xathin*

(1.2) **Example: cross–linguistic distribution of consonants and vowels in words**

a.	Hawaiian	allows no more than one consonant in a row	*wahine* 'woman' *alapine* 'often'
b.	English	allows long strings of consonants... but doesn't require them.	construct; sprig sɪksθs (six*ths*) *maven*
c.	Berber	allows words to consist solely of consonants... but also allows vowels in words.	*trglt* 'lock' *txdmt* 'gather wood' *ildi* 'pull'
d.	Yawelmani	allows up to two consonants in the middle of words... but allows at most one consonant at word edges.	*xathin* 'ate' *xathin* 'ate'

The four languages illustrated here demonstrate that there is a wide range of ways in which consonants and vowels distribute themselves within words in the world's languages. Significantly, there are also many patterns of consonants and vowels that you can think of that simply do not occur in natural languages. One such imaginable but non–occurring language would stack up all the consonants at the beginning of the word and all the vowels at the end of the word (1.3a). Words like *mrnaia* would exist, but no words like *marina*. A more "language–like" example would be comparable to English except that it *requires* all words to start with two or more consonants (1.3b). Words like *sprig* would be well–formed in this language, but not a word like *construct*, for *construct* begins with a single consonant.

(1.3) **Some imaginable but non–occurring languages**

a.	All consonants are in a sequence at the left edge of the word, followed by all vowels.	OK: not OK:	*spree, blue, mrnaia* *sprig, lube, marina*
b.	Every word begins with a string of consonants, otherwise like English.	OK: not OK:	*string, sprig, blue* *ring, pig, every*

There are no languages like those sketched in (1.3) and yet *it is not hard to describe such patterns*. In fact, many of the nonexistent patterns are *easier* to describe than some of the patterns found in natural languages, such as those in (1.2).

Through this extended discussion of consonant and vowel distribution, we have arrived at the central issues facing students of language. Although our example has been in terms of the sound systems of languages, the questions themselves are general and extend to all domains of language study.

(1.2d) is two sounds, as in *hot headed*; not one, as in *ether*. Finally, the sound symbolized by [x] in (1.2c, d) is a voiceless velar fricative, the final sound in the German pronunciation of a name like *Bach*. Following conventions of the field, square brackets are used to enclose symbols which represent sounds directly, such as [θ], [th], and [x] above.

> **The three central questions in linguistic research:**
> 1. What are the patterns that occur in natural languages?
> 2. How do we characterize the occurring patterns?
> 3. How do we exclude the patterns that we do not find and that we think we never will find?

In the next section, we explore the distribution of consonants and vowels in Yawelmani to illustrate the way linguists try to answer these three questions. We first examine the general answers to these questions, then turn to the Optimality Theoretic answers to questions of variation and universals to illustrate the workings of that model.

Yawelmani CV Distribution

There are four basic facts about the distribution of consonants and vowels in Yawelmani. First, words must begin with a consonant (1.4a). For example, the word *xathin* begins with a single [x]. By contrast, the Hawaiian word *alapine* 'often' starts with a vowel; a sequence like *alapine* is not a possible word in Yawelmani. In Yawelmani, a word like *alapine* would have the same "feel" to it as [bnɪk] *bnick* has in English: the sounds are all acceptable in the language but their organization is wrong. (Contrast this with the sequence [blɪk] *blick*, where the sounds and the organization are acceptable for English words, even though it happens not to be a word in English. See Chapter 2 for further discussion of words, possible nonwords, and impossible sequences.)

Second, at word edges, Yawelmani allows no more than one consonant (1.4b). This contrasts with words like [strɛŋθ] 'strength' in English where several consonants in sequence are found at both edges. (Note that Yawelmani words are not *required* to end in a consonant, unlike the word–initial requirement already discussed. However, if a Yawelmani word does end in a consonant, it may have only *one* consonant at the end.)

Third, there is no more than one vowel in a sequence in Yawelmani (1.4c). In *xathin*, both the *a* and the *i* are flanked by consonants. This contrasts with a language like Hawaiian or English, which both allow two vowels in a row. Finally, in the middle of words you can get at most two consonants together, although one consonant by itself is also permitted (1.4d). In this regard as well, Yawelmani differs from English, for English allows longer medial strings of consonants as in *persnickety*. (Examples in the "not OK" column are drawn from real words of Hawaiian, English, and Berber.)

(1.4)   **Additional Yawelmani facts**

		OK	NOT OK
a.	Words begin with exactly one consonant.	*xathin*	*a*loha 'greeting'; *a*pple, *o*dd
b.	At word edges, only one C is allowed.	*xathin*	*strength*
c.	No more than one vowel occurs in sequence.	*xathin*	k*ei*ki 'child', l*eo* 'voice'; al*ie*n (al*ie*n)
d.	Word–internally, CC is OK, but not necessary.	*xathin*, *xaten*	in*str*uct, co*nsp*ire; *trglt* 'lock'

As stated, the observations listed above do not reveal any obvious pattern. Other statements could be constructed, for example statements like those in (1.4) except replacing the word "consonant" by "vowel" and vice versa. That is, not only is the existence of a pattern unclear from the above, but so is its relation to universals governing the arrangement of consonants and vowels in words.

The statements in (1.4) are apparently unrelated observations about the placement of consonants and vowels in words. Although these locations can be stated clearly, it is not obvious from the list why these particular patterns are found and not others. In order to make sense of such observations about different languages, linguists have proposed that consonants and vowels are organized into constituents composed of consonants and vowels, called **syllables**, and that *words are composed of syllables*. That is, the distribution of consonants and vowels is characterized in terms of where each occurs in a syllable, a chunk smaller than a whole word and whose properties are easier to characterize. The distribution of consonants and vowels in words follows from the patterns that result when syllables are strung together.

In the next section, we examine the way in which phonologists characterize these facts to reveal the essential organization of consonants and vowels in Yawelmani words.

Words are Composed of Syllables

Under the assumption that words are composed of syllables, the linguist characterizes possible syllables, rather than possible words, both universally and for a given language. (See also Chapter 2.) In (1.5), I list certain general tendencies of syllables.

The terms used in the right–hand column of (1.5) are standard for referring to the parts of a syllable. A "CVC" syllable like [kæt] 'cat' has an **onset** [k], the initial consonant; it also has a **peak** [æ], the vowel; and it has a **coda** [t], the syllable–final consonant. A **complex onset** and a **complex coda** are found in [klæsp] 'clasp', which begins with two consonants, [kl], and ends with two, [sp]. The symbol "*" is used by linguists to indicate unacceptability. For example, placing a * at the beginning of a sentence indicates that the sentence is ungrammatical: *John seems that he ran*. Thus, *COMPLEX is shorthand for "complex onsets and complex codas are unacceptable".

(1.5) **Typical properties of syllables**

a.	Syllables begin with a consonant.	ONSET
b.	Syllables have one vowel.	PEAK
c.	Syllables end with a vowel.	NoCODA
d.	Syllables have at most one consonant at an edge.	*COMPLEX
e.	Syllables are composed of consonants and vowels.	ONSET & PEAK

There are two points of significance here. First, these statements are general tendencies, not absolute laws. Thus, there are syllables in languages which violate some of these properties, a point that OT exploits as we will see below. Second, the standard definition of a syllable, a constituent composed of at least one consonant followed by a vowel, results from combining (1.5a) and (1.5b): if a syllable starts with a consonant, it satisfies ONSET and if it has a vowel, it satisfies PEAK. This is one example of the observation that, by characterizing syllables in terms of the four simple properties in (1.5a–d) (which must be stated at any rate), further properties are also characterized.

In the next section, we see how each of these properties is manifested in Yawelmani. We also see how sequences of well–formed Yawelmani syllables result in the distribution of consonants and vowels in Yawelmani words, listed above in (1.4).

Explaining Yawelmani Consonant and Vowel Distribution Using Syllables

Figure (1.6) shows how the general tendencies of syllables (given in (1.5)) are realized in Yawelmani. The only one of these tendencies that does not hold absolutely in Yawelmani is NoCODA (1.6c) since some syllables do end with consonants.

(1.6)   **Properties of Yawelmani syllables**

		general tendency	Yawelmani
a.	PEAK	Syllables have one vowel.	always
b.	ONSET	Syllables begin with a consonant	always
c.	*COMPLEX	Syllables have at most one consonant at an edge.	always
d.	NoCODA	Syllables end with a vowel.	sometimes

The chart in (1.7) shows that if the only violable constraint in Yawelmani is NoCODA, then two types of syllables result, a CV syllable (1.7a) and a CVC syllable (1.7b). (We postpone discussion of how to characterize which constraints are violated in a particular language until Section 3.) Other imaginable syllable types, such as CVCC or CC (1.7c,d), are impossible in this language. CVCC syllables do occur in other languages, for instance in English *cart*, *desk*, and *tact*. English, then, tolerates violations of *COMPLEX. CC syllables occur in Berber; Berber tolerates violations of PEAK. A language which allows no violations of syllable constraints whatsoever has only CV syllables, (1.7a).

(1.7)   **How the Yawelmani syllable properties give rise to syllables**

		PEAK	ONSET	NoCODA	*COMPLEX
☞	CV	OK	OK	OK	OK
☞	CVC	OK	OK	FALSE	OK
*	CVCC	OK	OK	OK	FALSE
*	CC	FALSE	OK	OK	OK

The virtue of proposing that words are composed of syllables is that once we characterize the syllables of a language, lists of observations such (1.4) are seen to be exactly the properties we would expect. The discussion below shows how the observations about Yawelmani that are listed (1.4) are formally characterized by the properties given in (1.5), schematized below.

Yawelmani words begin with exactly one consonant (1.4a). Since each *syllable* necessarily begins with a consonant (ONSET), each *word* also begins with a consonant. Neither a vowel alone nor a vowel–consonant sequence is a syllable and so no word in the language can start with a vowel. By comparing the syllabification of *xathin* with that of the

English word *aha*, we see that violations of ONSET are not tolerated in Yawelmani, although they are allowed in English.

In (1.8), the violation of ONSET is indicated by an asterisk (*). The pictures here use a standard notation for syllables, where "σ" stands for "syllable" and syllable membership is shown by a triangle between a σ at the top and a sequence of consonants and vowels at the bottom. Another notation, used elsewhere in this chapter and book, uses dots to separate syllables: *xat.hin* and *a.ha*.

(1.8)   **Words begin with exactly one consonant: ONSET**

	ONSET
☞ σ σ △ △ xat hin	
σ σ ı △ a ha	*

not a possible word in Yawelmani

At word edges in Yawelmani, only one C is allowed (1.4b). The longest string of consonants that Yawelmani allows at the beginning of a word is the single consonant which satisfies ONSET. Similarly, a word might end with a single consonant, or with a vowel, because that is how syllables end. But a word cannot begin or end with two or more consonants. Our characterization of syllables explains this point: due to *COMPLEX, syllables do not end with more than one consonant. Since words are composed of syllables, words cannot end that way either.

(1.9)   **At word edges, only one C is allowed: *COMPLEX**

	*COMPLEX
☞ σ σ △ △ xat hin	
σ △ striŋ	*

not a possible word in Yawelmani

In Yawelmani, no more than one vowel occurs in a row (1.4c). With CV and CVC syllables, it is impossible to get two vowels in a row. Consonants must always intervene between vowels due to the necessary syllable–initial consonant. To get two vowels in a row, a syllable would have to be able to start with a vowel, a violation of ONSET, as shown in (1.10). In such a case, the second syllable is defective: unlike some languages (like English) in which syllables may start with vowels, no Yawelmani syllable does so.

In Yawelmani, word–internally, CC is OK (1.4d). The longest possible uninterrupted sequence of consonants in Yawelmani is two: this occurs if the syllable on the left ends with a consonant — recall that the one on the right must begin with a consonant. Since syllables cannot begin or end with a sequence of consonants, no more than two consonants in sequence arises from two syllables in sequence. Since syllables must have a peak (a vowel), the single C cannot be a syllable as in the third candidate in (1.11).

(1.10)   No more than one vowel occurs in sequence: ONSET

	ONSET
☞ σ σ △ △ xat hin	
σ σ △ △ xa in	*

not a possible word in Yawelmani

(1.11)   Word–internally, CC is OK: *COMPLEX and PEAK

	*COMPLEX	PEAK
☞ σ σ △ △ xat hin		
σ σ △ △ log whin	*	
σ σ σ △ │ △ log w hin		*

not a possible word in Yawelmani

not a possible word in Yawelmani

To summarize, we have made two proposals. First, words are composed of syllables. Second, syllables in Yawelmani are limited by the criteria in (1.6a–c). An immediate result of these proposals is that the list of facts about the distribution of consonants and vowels in Yawelmani words follows; nothing further need be said.

The role of syllable structure
1.   Words are composed of syllables.
2.   The facts about the distribution of consonants and vowels in a language follow from the structure of syllables in that language.

It is important to examine more closely the syllable properties given in (1.5) and (1.6). These sets of statements are stated as constraints on specific aspects of a syllable. Each of these statements expresses a strong universal tendency. For example, although it is not the case that all languages require onsets (ONSET), it is the case that *every language allows onsets* and *no language disallows onsets*. By allowing these constraints to be violated in some languages, two results are accomplished. First, language specific patterns and variation between languages are admitted into the model through such violations. Second, markedness is admitted into the model: each constraint violation indicates markedness in that respect. Employing constraints as we have done so far addresses the issues central to linguistic analysis: patterns, variation, universals, and markedness.

Constraints characterize universals.
Constraint violations characterize markedness, patterns, and variation.

3 Constraint Ranking and Faithfulness

Optimality Theory proposes that Universal Grammar contains a set of violable constraints. The constraints, as noted above, spell out universal properties of language. OT also proposes that each language has its own ranking for these constraints. Differences between constraint rankings result in different patterns, giving rise to systematic variation between languages.

> **Optimality Theory views...**
>
> Universal Grammar as a set of violable constraints
>
> the grammars of specific languages as the language–particular ranking of those constraints

The constraints include ones governing aspects of the phonology, such as the syllabification constraints just examined. The constraints also include ones governing morphology, including constraints which determine the appropriate position of morphemes (see Chapter 4). Finally, the constraint set includes constraints which determine the correct syntactic properties of a language, such as "a noun phrase must have case" (see Chapters 5 and 6). There is one family of constraints whose properties cut across all subdisciplinary domains, namely the **FAITHFULNESS constraints**, which say that the input and output are identical. Violations of FAITHFULNESS lead to differences between input and output, such as the difference seen in the prefix *in–/im–*, discussed in Section 1 above.

> FAITHFULNESS constraints require that the output be identical to the input.

Violation of constraints is tolerated in a very limited context. A constraint may be violated successfully only in order to satisfy a higher ranked constraint.

In this section, we examine how constraint ranking and violation handles the Yawelmani data. As already seen, in Yawelmani only one of the basic syllabification constraints may be violated, NOCODA. As a violable constraint, NOCODA must be outranked by some more important constraint(s). In this case, the relevant constraints are FAITHFULNESS constraints, one requiring faithfulness of consonants between input and output (FAITHC) and the other requiring faithfulness of vowels (FAITHV). In order to understand the necessary constraints, it is useful to explore the options available to avoid a NOCODA violation.

Syllabify the consonant as a peak. The first possibility is simply to syllabify the offending consonant as a syllable in and of itself, thereby violating PEAK. In this way, the offending consonant is now a peak, and so no longer violates NOCODA. Again, Yawelmani does not take this option: *xa.ten*, not **xa.te.n*, 'will eat'. (The dots refer to syllable boundaries, e.g. in *xa.ten*, the first syllable is *xa* and the second is *ten*, while in *xa.te.n*, there are three syllables, *xa*, *te*, and *n*. The * indicates that the syllabification *xa.te.n* is ill–formed.)

Delete the offending consonant. A second possibility is to simply delete the consonant which would otherwise be syllabified as a coda, thereby violating FAITHC. In Yawelmani, this does not happen. In the word *xaten* 'will eat', composed of *xat–* 'eat' and *–en* 'future tense', the word–final *–n* is also in coda position. It does not delete: *xa.ten*, not **xa.te*, 'will eat'.

Insert a vowel. A final possibility is to add a vowel after the offending consonant, constituting a violation of FAITHV. In this way, the offending consonant is now an onset for the added vowel, and so no longer violates NOCODA. Again, Yawelmani does not take this option: *xa.ten*, not **xa.te.ni*, 'will eat'.

The figure in (1.12) illustrates the above discussion in the manner common to most work in Optimality Theory. The figure is called a **tableau**; the constraints are ranked across the top, going from highest ranked on the left to lowest ranked on the right. Solid lines between constraints indicate crucial rankings while dashed lines indicate that the ranking is not (or not yet) crucial. In this example, for instance, it is crucial that NOCODA be subordinate to all other constraints. (Ultimately, we will see that FAITHV ranks below the other constraints and above NOCODA.)

The top left–hand cell shows the input representation for which candidates are being considered. Candidates show up in the leftmost column, with the optimal candidate indicated by the symbol "☞". The optimal candidate is the one with the fewest lowest violations. Violations are indicated by asterisks (*), and an exclamation point highlights each "fatal" violation, i.e. the violation that eliminates a candidate completely.[2] Shaded areas indicate constraints that are irrelevant due to the violation of a higher ranked constraint. (Only NOCODA gets shaded since it is the only constraint that must be dominated given the data considered so far.)

(1.12)   A tableau showing [xa.ten] is the optimal candidate

/xat–en/	PEAK	ONSET	*COMPLEX	FAITHC	FAITHV	NOCODA
☞ xa.ten						*
xa.te.n	*!					
xa.te				*!		
xa.te.ni					*!	

The first candidate in (1.12), [xa.ten], is the optimal one. Its only violation is that of NOCODA, the lowest ranked constraint. If the final consonant is syllabified as a peak, as in [xa.te.n] (the second candidate in (1.12)), the NOCODA violation is avoided, but only at the cost of a fatal PEAK violation.

The role of FAITHFULNESS becomes apparent when we consider the next two candidates, both of which avoid the NOCODA violation at the expense of a FAITH violation.

---

[2]Given that GEN creates an infinite set of candidates, a necessary strategy when presenting tableaux is to restrict the candidates presented in the tableau to those which are critical to the point being made — the infinite set could not possibly be considered! Thus, the candidates in (1.12) are limited to four relevant ones. A second, similar, strategy is to omit from tableaux the candidates which violate undominated constraints in the language. The tableau in (1.12) contains an example: since it has been established that Yawelmani never violates ONSET, [xat.en] (with an ONSET violation) need not be included in subsequent tableaux.

Candidate [xa.te] in (the third candidate in (1.12)) has lost the final consonant thereby incurring a FAITHC violation. Since FAITHC is higher–ranked than NOCODA, this violation is fatal, and eliminates *xa.te* from consideration. The form [xa.te.ni] in (the fourth candidate in (1.12)) suffers a comparable fate. In this form, an extra vowel allows the last consonant to syllabify as an onset, which eliminates the NOCODA violation. Due to the extra vowel, however, the form incurs a fatal violation of FAITHV.

> Satisfying some higher ranked FAITHFULNESS constraint (s) in Yawelmani may force the violation of lower ranked constraint(s).

4 Optimality Theory

At this point, we have worked through a specific example in order to understand the way in which Optimality Theory works. In this section, I take a more formal approach in laying out properties of the model.

Optimality Theory, like other models of linguistics, proposes an input and an output and a relation between the two. In transformational (or derivational) views, which have been the dominant paradigm in linguistic research since the mid 1960s, the input is the starting point, there is a series of operations performed on the input, and the result of these operations is the output. Crucially, if an operation makes some change in the input, that changed form serves as the input to the next operation. See especially Chapters 3 and 4 for discussion of this point.

In OT, the relation between input and output is mediated by two formal mechanisms, **GEN** and **EVAL**. GEN (for **Generator**) creates linguistic objects and notes their faithfulness relations to the input under consideration. EVAL (for **Evaluator**) uses the language's constraint hierarchy to select the best candidate(s) for a given input from among the candidates produced by GEN. The constraint hierarchy for a language is its own particular ranking of **CON**, the **universal set of constraints**.

The roles of GEN, EVAL, and CON are illustrated in (1.13), which schematically presents how OT determines the optimal syllabification for the input /xat–en/. The input feeds into GEN, which creates candidates. The candidates are considered by EVAL, which selects the optimal candidate from the set.

This specific example is the same one we have just worked through, namely determining the optimal syllabification for /xat–en/. The tableau in (1.12) makes explicit the role of EVAL. However, as can be seen by inspecting (1.13), there is quite a bit more to the formal aspects of OT than simply reading a tableau. In this section, we examine the Input, GEN, CON, and EVAL.

The Input

Universal Grammar provides a vocabulary for language representation; all inputs are composed from this vocabulary. As a result, inputs are linguistically well–formed objects in the sense that the input does not contain nonlinguistic objects. This is the sole restric-

14

DIANA ARCHANGELI

tion imposed on the input: all other constraints are found in EVAL. In the specific example we have been examining, the vocabulary provided by Universal Grammar must include consonants, vowels, and syllables. For other examples discussed in this book, this vocabulary must include phonological features and the categories *noun*, *verb*, etc.

The Formal Model

1. **GEN** — for a given input, the **Generator** creates a candidate set of potential outputs
2. **EVAL** — from the candidate set, the **Evaluator** selects the best (optimal) output for that input
3. **CON** — EVAL uses the language particular ranking of constraints from the **universal set of constraints**

(1.13) **A schematic of OT**

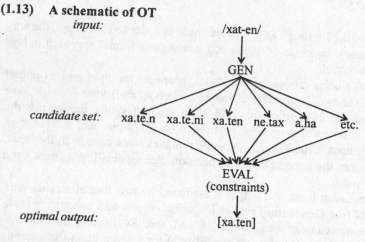

GENerator

GEN is a function which relates the input to a set of candidate representations, any one of which may be the optimal output form for the specific input. GEN is restricted in that it can only generate linguistic objects, ones composed from the universal vocabulary that similarly restricts inputs.

GEN is quite creative, being able to add, delete, and rearrange things without restriction. Since there are no restrictions, the candidate set created by GEN for any given input is infinite. (One candidate has one added element, another has two added elements, etc.) This particular property is a serious problem for those who wish to implement Optimality Theory either as a production and processing model or as a computational model, although efforts have been made to surmount the challenges.

GEN also has the job of indicating correspondences between input and output representations. These correspondences are crucial in evaluating the FAITHFULNESS constraints, such as FAITHV and FAITHC. How correspondence is encoded by GEN and how it is treated by EVAL is a subject of controversy in current Optimality research; however, all researchers agree that such correspondences must be encoded in some way. See Chapter 4 for relevant discussion.

The Universal CONstraint Set

CON, as a universal set of constraints, is posited to be part of our innate knowledge of language. What this means is that every language makes use of the same set of constraints. This assumption leads directly to a characterization of the universal aspects of human language: all languages have access to exactly the same set of constraints. This is the formal means by which *universals* are encoded.[3]

Whether stated in a positive fashion (e.g. ONSET) or negatively (e.g. *COMPLEX), any constraint may end up being violated in some language: the potential for being violated is a result of the position of a constraint in a particular language's hierarchy, rather than a property of the constraint itself. In this way, the constraints also provide a measure for markedness: the higher ranked constraints (and so rarely violated) indicate the ways in which the language is unmarked while the lower ranked (and so frequently violated) constraints indicate the ways in which the language is marked. As such, markedness is encoded directly in the model. This is an important result, for earlier models have required separate theories of markedness.

EVALuation

EVAL is the mechanism which selects the optimal candidate (s) from the candidate set created by GEN. EVAL makes use of a ranking of the violable constraints. The optimal output, the one that is selected by EVAL, is the one that best satisfies these constraints.

EVAL is at the heart of Optimality Theory
1.   The constraints in CON are violable.
2.   The constraints are ranked.
3.   EVAL finds the candidate that *best satisfies* the ranked constraints.
    a.   Violation of a lower ranked constraint may be tolerated in order to satisfy a higher ranked constraint.
    b.   Ties (by violation or by satisfaction) of a higher ranked constraint are resolved by a lower ranked constraint.

Best satisfaction can be achieved in two ways. Violations of lower ranked constraints are tolerated in the optimal form provided that they help avoid violation of some higher ranked constraint. Lower ranked constraints adjudicate when all viable candidates tie on

---

[3]The ideal which Optimality research aims for (and sometimes appears to fall short of) is to provide evidence of the universality of each constraint necessary for some particular language. For constraints such as the ones posited for syllabification in this chapter, universality is readily motivated; there are numerous analyses involving constraints whose status as a universal is minimal at best. At this point, it is unclear whether this is a weakness of the model itself, or a weakness of the analyses.

some higher ranked constraint, either because all candidates satisfy the higher constraint or, more interestingly, because all candidates violate the higher ranked constraint. In this way, unmarked patterns that are encoded in low–ranked constraints occasionally emerge despite those patterns not being observed throughout the language. See especially McCarthy & Prince (1994).

Summary

In this section, we have briefly explored the basic parts that Optimality Theory proposes for grammars, input, GEN, EVAL, and CON. We have seen two ways in which *universal* properties of language are encoded: (i) inputs and GEN are limited by a universal linguistic vocabulary; (ii) CON contains a universal set of constraints which all languages use. These properties are illustrated with the Yawelmani example already discussed: ONSET, PEAK, and *COMPLEX are all considered very strong universal properties.

How Optimality Theory works

1. *Universal Grammar includes*
     a.   a linguistic alphabet
     b.   a set of constraints, **CON**
     c.   two functions, **GEN**(erate) and **EVAL**(uate).
2. *The grammar of a particular language includes*
     a.   basic forms for morphemes (from which inputs are constructed)
     b.   a **ranking** for the constraints in **CON**
3. *For each input,*
     a.   **GEN** creates a candidate set of potential outputs
     b.   **EVAL** selects the optimal candidate from that set

---

We have seen that *markedness* is encoded via constraints and constraint violations. Yawelmani is unmarked with respect to ONSET, PEAK, and *COMPLEX, since these constraints are never violated. However, since some Yawelmani syllables end with consonants, NOCODA can be violated. Yawelmani is marked in this respect. Under OT, markedness arises when a constraint, such as NOCODA, is violated: such violation occurs only in order to satisfy higher ranked constraints.

The *patterns* found in languages are characterized by the language–specific ranking of the universal constraints. This connects closely to language *variation*, which also arises from the different ways in which languages rank the constraints in CON. This characterization of patterns and variation is one of the most exciting and intriguing aspects of OT. The next section provides an illustration, comparing the effects of different rankings of *COMPLEX, PEAK, FAITHC, and FAITHV.

> **How Optimality Theory addresses the issues that concern linguists.**
> 1. **Language variation** is characterized as different rankings of the same set of constraints.
> 2. **Specific patterns** are derived from the language–particular rankings of these constraints.
> 3. **Universals** are present in the universal — but violable — constraints.
> 4. **Markedness** is inherent in the model.
>   a. Each constraint is a markedness statement.
>   b. Specific aspects of markedness result from the ranking.

5 Language Variation as Differences in Constraint Rankings

Language variation, as already noted, also follows from the role of the constraints within particular languages. Two constraints A and B may be ranked A » B in one language and B » A in another. Each ranking characterizes the distinctive patterns of the two languages and leads to variation between them.

To illustrate language variation via differences in constraint rankings, we return to the Yawelmani data and examine some forms related to the two we have already considered, *xathin* 'ate' and *xaten* 'will eat'. The additional data supports the partial constraint ranking FAITHC, PEAK, *COMPLEX » FAITHV. We then consider alternative rankings of these four constraints, and see that different rankings result in different patterns of syllabification. Spanish, English, and Berber provide examples.

Additional Yawelmani Data

As already seen with *xaten*, adding *–en* to the end of a verb marks the future tense. Given that *logwen* means 'will pulverize', the verb root for 'pulverize' must be *logw–*, as shown in (1.14b).

(1.14) **Yawelmani future tense: a V–initial suffix**

	word	morphemes	gloss
a.	xaten	xat–en	'will eat'
b.	logwen	logw–en	'will pulverize'

Stripping off the *–en* suffix reveals the bare verb roots, *xat–* 'eat' and *logw–* 'pulverize'. Significantly, the root *logw–* is an unsyllabifiable sequence in Yawelmani because it ends with two consonants. Because the future suffix *–en* begins with a vowel, the consonant sequence can safely syllabify: the /g/ is the coda of the first syllable and the /w/ the onset of the second syllable, (1.15b). However, this type of root presents a problem if there are any consonant–initial suffixes: attaching a suffix which begins with a consonant to a verb root like *logw–* will necessarily create a sequence of *three* consonants. As we have already seen, no more than *two* consonants occur in a sequence in the words of this language.

(1.15) Syllabification of *xaten* and *logwen*
  a.   xa.ten   b.   log.wen

In fact, Yawelmani does have consonant–initial suffixes, for example, the past tense suffix –*hin*. When –*hin* is added to the verb root *xat*–, the two consonant sequence *th* is created. These two consonants can each find a position in a syllable, the *t* is the coda of the first syllable while the *h* is the onset of the second syllable.

(1.16) Yawelmani past tense: a C–initial suffix

word	*morphemes*	*gloss*	*syllabification*
xathin	xat–hin	'ate'	xat.hin

There are four logically possible alternatives, illustrated in (1.17). The first three solutions to the problem have outputs that are segmentally faithful to the input but have odd syllables. A complex coda (1.17a–i) or a complex onset (1.17a–ii) might be created, or the extra consonant might simply form a syllable by itself, (1.17b). The remaining two solutions each involve a mismatch between the input and the output: something is either added or lost. In both cases the resulting syllables are well–formed. First, the extra consonant might be deleted, indicated by the ᴡ in (1.17c). Second, a vowel might be added, (1.17d). Adding a vowel allows the extra consonant to syllabify without creating a complex onset or coda.

(1.17) Syllable well–formedness and the extra consonant

  a. i.   *create a CVCC syllable*   a. ii.   *create a CCVC syllable*

logwhin   logwhin

  b.   *create a C syllable*

logwhin

  c.   *lose a consonant*   d.   *add a vowel*

logᴡhin   logiwhin

Figure (1.18) shows how the relevant constraints considered thus far stack up with respect to the logical possibilities given in (1.17). The chart shows that whenever an input representation has a three consonant cluster, one of these constraints must be violated. The question is "what do you do?" The very interesting answer is that what you do depends on what language you speak. There are languages of each type.

The chart in (1.18) shows that when a form is completely faithful (i.e. there is nothing added or deleted), one of two bad results occurs: either a violation of *COMPLEX (the first two candidates in (1.18)) or a violation of PEAK (the third candidate in (1.18)). The unfaithful candidates avoid those two violations in one of two ways: by deleting a consonant and so incurring a FAITHC violation (the fourth candidate in (1.18)), or by adding a vowel, at the cost of violating FAITHV (the fifth candidate in (1.18)).

(1.18)   **An input /...CCC.../ sequence**

/logw–hin/		*COMPLEX	PEAK	FAITHC	FAITHV
logw.hin	BAD	*			
log.whin	BAD	*			
log.w.hin	BAD		*		
log._hin	BAD			*	
lo.giw.hin	BAD				*

> Every output resulting from an input /...CCC.../ sequence must violate some constraint!

FAITHV is Outranked in Yawelmani

In Yawelmani, we discover that a vowel, [i], is added to rescue an otherwise unsyllabifiable consonant, a signature property of this language. We have already seen this pattern with *logwen* 'will pulverize' and *logiwhin* 'pulverized'; (1.19) provides an additional example, with the verb /ʔilk–/ 'sing'.[4] In all pairs of Yawelmani words related by the presence or absence of some vowel, the vowel [i] is always the relevant one.

(1.19)   **Words related by the presence or absence of [i]**

	attested word	not...	gloss
a.	logiwhin	*logwhin, *loghin	'pulverized'
	logwen	*logiwen	'will pulverize'
b.	ʔilikhin	*ʔilkhin, *ʔilhin	'sang'
	ʔilken	*ʔiliken	'will sing'

Once we accept that roots like *logw–* 'pulverize' and *ʔilk–* 'sing' end with two consonants, we must explain why the vowel [i] is inserted in certain forms but not in others. The short answer is that the vowel is added in order to allow an extra consonant to be syllabified (1.17d). This prevents the loss of a consonant, shown (1.17c) as well as the ill-formed syllables, as in (1.17a,b).

The challenge when addressing such facts is to formally characterize the relation between the input *logw–hin* and the output *logiwhin* in a manner that expresses both universals and variation. The Optimality Theory response is that such patterns are stated in terms of constraint satisfaction and violation, where the constraints themselves express the universals and the particular constraint rankings express variation.

---

[4]The symbol [ʔ] is a glottal stop. This sound is produced by some speakers of English between the two words in "a apple" (not "*an* apple") and in some dialects of English in place of the "t" in words like *bottle*.

The option preferred in Yawelmani is to insert a vowel, thereby saving all consonants and avoiding complex syllable margins (onsets and codas). The first two forms in (1.20) have retained all consonants by putting two at the edge of one of the syllables, either [gw] closing the first syllable or [wh] starting the final syllable, fatally violating *COMPLEX. The third form in (1.20) is completely faithful in terms of segments, but has syllabified a consonant as the peak of a syllable, thereby violating the high–ranked PEAK. The fourth form in (1.20), has lost one of the three consonants and so violates FAITHC, also a fatal violation due to its high ranking.

(1.20)   **Yawelmani: *COMPLEX, FAITHC, PEAK » FAITHV**

/logw–hin/	*COMPLEX	FAITHC	PEAK	FAITHV
logw.hin	*!			
log.whin	*!			
log.w.hin			*!	
log . hin		*!		
☞ lo.giw.hin				*

The final form in (1.20) includes an extra vowel, which allows syllabification of the third consonant, but which violates FAITHV. By hypothesis, though, FAITHV is the lowest ranked of these constraints in Yawelmani: for the input /logw–hin/, [logiwhin] is selected as the optimal output form.

Three other rankings remain to be considered, those with each of FAITHC, *COMPLEX, and PEAK as the subordinate constraint. Although a formal possibility with constraint ranking is that these four constraints are crucially ranked with each other (giving 4!, or 24 possible rankings) we need not consider all of the logically possible rankings. Constraint rankings are crucial only when they decide between competing candidates, but the data and constraints under consideration here do not have that level of complexity. (On this point, see the discussion of syllable typology in Chapter 2.)

FAITHC is Violated in Spanish

The next example we consider illustrates the "lose–a–consonant" option. In such cases, FAITHC is subordinate: in a three consonant sequence, the best thing to do is to leave a consonant out. In this way, PEAK, FAITHV and *COMPLEX are satisfied, at the cost of violating FAITHC. An example is found in Spanish. A caveat is in order here. Even a small amount of familiarity with Spanish will reveal that the facts presented here are simplified somewhat. In particular, I ignore the well–known fact that Spanish inserts a vowel in front of *sC* clusters (e.g. *esfera* 'sphere'; compare *hemisferio* 'hemisphere', not **hemiesferio*.)

The first column in (1.21) shows Spanish verbs in the infinitive form. Each of the verb roots ends with two consonants when followed by an infinitive suffix, *–er* or *–ir*. for example, in (1.21a) the verb root ends with the two consonant sequence *–rb–*: *absorber*, and in (1.21b) it ends with the two consonant sequence *–lp–*: *esculpir*.

(1.21)   **Spanish data**

	infinitive	*adjective or noun*	
a.	absorb–er	absor–to	*absorbto
b.	esculp–ir	escul–tor	*esculptor
c.	distingu–ir	distin–to	*distingto

However, as the second column shows, where an adjective–forming suffix –to is added, the verb root appears to end with a single consonant. For instance, in (1.21a) the verb root ends with the single consonant –r–: absorto. What happened to the –b–? We can ask similar questions for (1.21b,c): what happened to the –p– and the –g–?[5] The answer is clear if we focus on the phonological shape of the suffixes, rather than their morphological function. The infinitive suffix is vowel initial. As such, the final consonant of the verb root (–b–, –p–, –g–) is syllabified as the onset for that vowel's syllable (1.22a). By contrast, the noun– and adjective–forming suffixes begin with consonants. The final consonant in the verb root cannot syllabify without creating a CC onset or a CC coda, thereby violating *COMPLEX, (1.22c).

(1.22)   **Syllabification of *absorber* and *absorto***

	a.		b.		c.	
	ab.sor.ber		ab.sor.to		*ab.sorb.to	*ab.sor.bto
	es.cul.pir		es.cul.tor		*es.culp.tor	*es.cul.ptor
	dis.tin.[gi]r		dis.tin.to		*dis.ting.to	*dis.tin.gto

Spanish apparently does not allow *COMPLEX violations. In this way, Spanish is like Yawelmani. Spanish does not allow a consonant to syllabify by itself, giving *[ab.sor.b.to], again like Yawelmani. Nor does it adopt the Yawelmani option, of inserting a vowel, resulting in *[ab.so.reb.to] or *[ab.sor.be.tʊ].

Significantly, the two languages differ in their resolutions to the "extra consonant" problem. As already seen, in Yawelmani the added vowel [i] rescues the unsyllabifiable extra consonant. In Spanish, however, the unsyllabifiable consonant is not rescued: it is simply deleted.

The tableau in (1.23) shows that ranking FAITHV, PEAK, and *COMPLEX above FAITHC results in consonant deletion. The first form in (1.23), which surfaces as [absorto] since the –b– is not syllabified, is the correct form. All the input vowels surface and no vowels are inserted, satisfying FAITHV. There are no CC onsets or codas, satisfying *COMPLEX. There are no syllables composed solely of consonants, satisfying PEAK. These results are achieved at cost, however: the root–final consonant –b– is not syllabified, incurring a violation of FAITHC. Due to ranking the other constraints above FAITHC, however, the FAITHC violation is not fatal.

This contrasts with the results in the second, third and fourth forms in (1.23), each of which incurs a fatal violation of some constraint ranked above FAITHC. In the second candidate in (1.23), FAITHC is satisfied because the –b– is syllabified. However, satisfaction of FAITHC comes at a perilous cost, the violation of *COMPLEX. Similarly, in the third candidate in (1.23), the –b– is again syllabified, this time as the onset to the syllable

---

[5]In Spanish orthography, a *u* follows the *g* in *distinguir* to indicate that the *g* is "hard", that is, pronounced like the *g* in English *get* and *Abigail*.

of an inserted vowel [absor<u>be</u>to].[6] Satisfaction of FAITHC again comes at cost, here a
violation of FAITHV due to the inserted vowel. Finally, in the fourth candidate in (1.23)
satisfaction of FAITHC is attained through violation of PEAK. By ranking FAITHC below
the other constraints, the correct form is selected.

(1.23)   Spanish: FAITHV, PEAK, *COMPLEX » FAITHC

/absorb–to/	FAITHV	PEAK	*COMPLEX	FAITHC
☞ ab.sor . to				*
ab.sorb.to			*!	
ab.sor.be.to	*!			
ab.sor.b.to		*!		

In both Yawelmani and Spanish, *COMPLEX and PEAK are high–ranked. The languages
differ in which FAITHFULNESS constraint is more critical, FAITHV or FAITHC. In Spanish,
FAITHV is more important: no vowel may be inserted to rescue the extra consonant. The
consonant is simply lost. In Yawelmani, FAITHC is more important: violating FAITHV is
countenanced as long as the effect of the violation protects a consonant from deletion.

We now turn to an example of a third type of language, one in which FAITHFULNESS is
more important than keeping syllables simple. Our example is English.

*COMPLEX is Violated in English

In English, we add a suffix –ness to adjectives in order to create nouns: happy,
happiness; sad, sadness; etc. Since –ness begins with a consonant, the critical environment is found when –ness is added to an adjective that ends in two consonants, such as
limp. Rather than limness, with consonant loss, or limp[i]ness, with vowel addition,
English creates a complex coda, resulting in limpness. In fast and/or casual speech, people may omit the t in the cluster ...ftn... and say so[fn]ess instead of so[ftn]ess. How best
to account for this has yet to be resolved satisfactorily. One possibility within OT is variable ranking of constraints, or variable values in constraints depending on speech
rate/style. Another possibility, discussed in Chapter 2, is the use of correspondence constraints between careful and casual speech representations.

The forms in the noun column are formed by simply adding –ness to the adjectives,
and syllabifying, without deleting or adding anything, exactly the pattern expected if
other constraints outrank *COMPLEX. This is illustrated by the tableau in (1.25). The cor-

---

[6]The alert reader will have noticed that I have changed the vowel which is inserted: in the
Yawelmani example, the vowel [i] is inserted while in Spanish the vowel is [e]. In the next
example, from English, the inserted vowel is [ɨ]. Why are the vowels different? The answer is that
this is yet another type of variation found between languages. Vowel qualities of inserted vowels
vary across languages. The vowel [i] is the vowel that is inserted in the Yawelmani example, while
in Spanish the vowel [e] is sometimes inserted (as in the esfera 'sphere' example), although it is
not inserted in the environment under discussion. In English, the vowel [ɨ] is inserted in certain
contexts: when the past tense or plural suffix adds a syllable to the word as in kisses [kɪsɨs] and
hinted [hɪntɨd]. (See Chapter 4.) Under Optimality Theory, the quality of the vowel is also
determined by constraint ranking, a complexity that is ignored in our discussion.

rect form, *limpness* is the first candidate in (1.25). There are no violations to PEAK, FAITHV or FAITHC; the sole violation is to the subordinate *COMPLEX. (In fact, the very existence of words like *limp*, *soft*, and *crisp* attest to the low ranking of *COMPLEX in English.)

(1.24)   **English data**

	adjective	noun		
a.	limp	limpness	*lim_ness	*limp[ɨ]ness
b.	soft	softness	*sof_ness	*soft[ɨ]ness
c.	crisp	crispness	*cris_ness	*crisp[ɨ]ness
d.	strange	strangeness	*stran_ness	*strang[ɨ]ness

(1.25)   **English: FAITHV, FAITHC, PEAK » *COMPLEX**

/lɪmp–nes/	FAITHV	PEAK	FAITHC	*COMPLEX
☞ lɪmp.nes				*
lɪm.nes			*!	
lɪm.pɨ.nes	*!			
lɪm.p.nes		*!		

Each of the failing candidates violates one of the other three constraints. The adjective–final consonant –p– is lost in the second candidate in (1.25), incurring a FAITHC violation in order to achieve universally better syllables. Similarly, the added vowel in the third candidate in (1.25) results in universally better syllables, but only at the cost of a FAITHV violation. Finally, the fourth candidate in (1.25) shows the extra consonant syllabified by itself, incurring a PEAK violation. The English pattern is characterized by ranking *COMPLEX below the other three constraints, the third of the four rankings we consider.

PEAK is Outranked in Berber

A striking fact about Berber is the long sequences of consonants that surface. The Berber data is taken from Dell & Elmedlaoui (1985), a discussion of the Imdlawn Tashlhiyt dialect. See Prince & Smolensky (1993) for a complete reanalysis of the data in terms of Optimality Theory.

A striking fact about Berber words is that they do not even need to contain vowels: *trglt* 'lock', *txdmt* 'gather wood', *trkst* 'hide', all in the second person singular perfective (e.g. 'you have locked'). The question, of course, is how such sequences are arranged into syllables. Berber accomplishes this via PEAK violations.

The chart in (1.26) contrasts two morphological categories, the third person masculine singular which consists of a vowel prefix *i–* and the third person feminine singular, a consonant prefix *t–*. When *i–* is added to a verb which starts with two consonants, nothing surprising happens: the first consonant closes the first syllable while the second is the onset for the second syllable, as in *il.di* 'pull'. The surprise occurs when the consonantal prefix *t–* is added: in this case, the verb's initial consonant syllabifies as the peak of a syllable: *tl.di* 'pull'. (A comparable situation is found in English, although it is perhaps not nearly so dramatic as the Berber pattern. For example, in *cylinder*, the final consonant, *r* serves as the peak of the final syllable while it is clearly in the onset of its syllable in the related word *cylindrical*.)

The pattern is exactly the one we expect if PEAK is outranked by the other constraints. In this event, the best solution to the problem of "too many consonants, not enough vowels" is to allow one of the consonants to assume the syllabic position normally reserved for vowels.

(1.26) **Berber data**

	3 m sg	*3 f sg*	*gloss*
a.	iz.di	tz.di	'put together'
b.	if.si	tf.si	'untie'
c.	is.ti	ts.ti	'select'

As illustrated by (1.27), each of the alternative syllabifications results in the violation of a higher ranked constraint. Creating a single syllable with the prefix violates *COMPLEX (the first candidate in (1.27)); deleting consonants violates FAITHC (the second, third and fourth candidates in (1.27)); and adding a vowel violates FAITHV (the fifth candidate in (1.27)).

(1.27) **Berber: FAITHV, FAITHC, *COMPLEX » PEAK**

/t–fsi/	FAITHV	*COMPLEX	FAITHC	PEAK
t.fsi		*!		*
_.si			*!*	
_f.si			*!*	*
t_.si			*!*	*
tif.si	*!			
☞ tf.si				*

Berber resolves the extra consonant problem by allowing consonants to syllabify as syllable peaks, a position that most languages reserve for vowels.

Summary

In this section, we explored the "extra consonant" problem: syllabification of a sequence of three consonants *must* violate one of four different constraints. Different rankings of the four constraints predict four different resolutions to the problem, depending on which constraint is lowest ranked. The syllabification patterns of four languages, Yawelmani, Spanish, English, and Berber, are exactly the four patterns predicted by the model.

6 Conclusion

This chapter began with a very brief introduction to linguistics, and a slightly longer introduction to phonology (Sections 1 and 2 respectively). We then explored the structure of Optimality Theory. Section 3 sketched an analysis of Yawelmani syllabification within OT and introduced the "tableau", an expository device used to demonstrate the effect of EVAL.

Section 4 provided a more formal discussion of the components of OT, including the input, GEN, EVAL, CON, and the output. A schematic summary picture is given below of the functions in a grammar under OT. The subscript "L" on the function EVAL indicates that EVAL is a *language–particular* ranking of CON, the universal set of constraints.

(1.28)   **Optimality Theory (Prince & Smolensky 1993)**
         GEN (Input$_k$) → {Candidate$_1$, Candidate$_2$, Candidate$_3$,...}
         EVAL$_L$ {Candidate$_1$, Candidate$_2$, Candidate$_3$,...} → Output$_k$

We then considered how Optimality Theory accounts for the central issues in linguistics: *universals*, *markedness*, patterns and *variation*. Universals are represented in OT by CON, the universal set of constraints. Markedness is represented in OT by constraint violation while constraint satisfaction corresponds to unmarked properties. Patterns are the result of the interplay between a particular constraint hierarchy and the inputs provided by the language. Variation results from differences in the constraint rankings selected by specific languages, illustrated in Section 5. Each language deals with ...CCC... sequences in a different way, and each way is characterized by a distinct ranking of the relevant constraints.

In addition to knowing how OT works and how OT accounts for the central linguistic issues, it is also useful and interesting to examine some other aspects of the model. Why has the model become so popular so fast? What areas of study are particularly suited to OT analysis? Does the model change what we thought we knew about language? What challenges does OT face?

In this closing section, I explore why OT caught on so rapidly, to the point that six years after its inception, it is the dominant paradigm in formal phonology, and is rapidly gaining ground in both morphological and syntactic analysis. I then consider two types of issues that remain to be explored, those concerned with the nature of the theory itself, and those addressing different empirical domains that may be amenable to OT analysis.

The Rise of OT

To understand the rapid and widespread acceptance of Optimality Theory, one must understand the state of theoretical research in linguistics in the late 1980s. In many ways, it was foundering. Consider phonology. Great advances in our understanding of representations were made throughout the late 1970s and continuing into the early 1980s, resulting in the nonlinear representations that are now widely assumed. There was great hope that as our understanding of representations improved, the characterization of alternations would be simplified. This simplification did not happen.

Efforts were also directed specifically at formally restricting the possible types of alternations. Efforts in this domain, too, were unsuccessful: the alternations permitted by every formal model unfortunately also include alternations that are both unattested and thought to be unlikely. There were always counterexamples.

Finally, constraints were being used — or rather, over–used. The standard of the underlying–surface relation included (i) the abstract underlying form of the morphemes of a language (ii) which concatenate and then (iii) undergo a series of rules. When no further rules apply, (iv) the surface form has been attained. (See also Chapters 3 and 4.)

(1.29)   **Generative Phonology**
                    underlying representation
                              ⇩
                    morpheme concatenation
                              ⇩
                          rules
                              ⇩
                    surface representation

This picture looks neat and tidy — until the role of constraints is added. Constraints entered this picture at all stages. Constraints hold of the underlying forms, in terms of what sounds are permitted and what sequences of sounds are licit. Constraints hold of morpheme concatenation, restricting how morphemes may combine. Constraints hold of rule application, limiting both how rules can apply and what types of sounds or sound sequences can be produced. Constraints hold of outputs, prohibiting patterns that do not occur at the surface. Unlike in OT, all of these types of constraints have been viewed as *inviolable* within the relevant domain.

(1.30)   **Phonology in the 1970s and 1980s**
          constraints hold here ⇨   underlying representation
                                                      ⇩
          constraints hold here ⇨   morpheme concatenation
                                                      ⇩
          constraints hold here ⇨   rules
                                                      ⇩
          constraints hold here ⇨   surface representation

The frustrations in syntax bear some similarities, but also some differences. Syntactic representations, too, have evolved into increasingly elaborate hierarchical structures which in turn have required the increasing use of empty terminal nodes. Inviolable constraints, generally called **conditions** or **principles**, have played a more dominant role in syntax than in phonology however, for syntacticians to a greater extent than phonologists have attempted to minimize the rule component.

The "inviolable" principles of syntax have themselves proved to be problematic in that inviolability has been purchased at the cost of a variety of types of hedges. As shown in detail in Chapter 6, some principles are "parameterized", holding in one way in one language and in another way in another language. Other principles have peculiar restrictions built–in. For example, the Extended Projection Principle begins with the strong claim "All clauses must have a subject", such as *John* in *John ran* and *it* in *It rained all night*. However, this principle is weakened by the codicil "*unless the language lacks expletives*", in order to account for subjectless sentences in languages like Yaqui such as *Yooko yukne* 'It will rain tomorrow' (literally 'Will rain tomorrow'; see Chapter 6, Section 2 for more discussion of this point.)

In both areas, research results have indicated that the general analytic strategy has been on the right track; at the same time, there had been growing dissatisfaction in two ways. First, despite continued innovations in theories of rules and of representations, certain types of data remained unexplained. Second, the prevailing belief about constraints — that they are inviolable — resulted in a continuing frustration with their role in grammar, for it is exceedingly difficult to find a constraint that is never violated.

Optimality Theory redefines the role of constraints and in so doing redefines the research focus. All constraints are violable. Grammars define the relative significance of violating specific constraints. Constraints are present only in the constraint hierarchy: there are no separate constraints on inputs nor on outputs. There are two powerful implications for linguistic analysis here. First, there simply is no rule component at all. Second, the constraint hierarchy must be constructed to return some result regardless of the input (the result may be a **null parse**, that is nothing at all). Examining the nature of underlying representations and of rules has been core to linguistic research: OT changes the focus of this research.

A desirable aspect of this change is that research focuses directly on universal properties of language: since the constraints are hypothesized to be universal, OT redirects our research focus towards language universals. This aspect of grammar must be central to any OT analysis, regardless of how language–specific the phenomenon is. OT has not yet been able to answer all the unanswered questions; however, it has provided a dramatically different approach to accounting for both universals and variability and to the input–output relationship.

Optimality Theory addressed these problematic issues.

1. **It defines a clear and limited role for constraints.**
   a. Each constraint is universal.
   b. Constraints are ranked in EVAL.
2. **It eliminates the rule component entirely.**
   Different constraint rankings in EVAL express language variability.
3. **It focuses research directly on language universals.**
   Each constraint is universal.
4. **It resolves the "nonuniversality of universals" problem.**
   Universals don't play the same role in every language.

Remaining Issues

Issues abound in each of the components of the OT model. OT challenges the way in which we think about linguistic representations and relations such that virtually every aspect of previously held assumptions must be reconsidered.

The input. Linguists are only beginning to explore the nature of the input under OT. There are four classes of problems here. First is the issue of the lack of constraints on the input, an aspect of the theory known as Richness of the Base. In standard generative phonology, numerous constraints were imposed on the input. For example, an analysis would frequently begin by defining the vowel and consonant inventories of a language in terms of permitted feature combinations. See Chapter 3 for the analysis of inventories in OT. Under OT, such constraints on the sounds that make up the input are impossible. Inputs are potentially as infinite as the candidate set; the constraints in EVAL must be ranked so that impossible sounds or sound sequences never surface.

Since there are no constraints on the input, it is easy to construct multiple inputs that converge on a single output. Which of the multiple inputs is the best one? A variety of strategies are imaginable. For example, an algorithm called **lexicon optimization** is introduced in Prince & Smolensky (1993) and developed in Itô, Mester, & Padgett (1995). Lexicon optimization examines the constraint violations incurred by the winning output candidate corresponding to each competing input. The input–output pair which incurs the fewest violations is considered the optimal pair, thereby identifying an input from the output.

Second is the issue of what exactly goes into an input? Most people assume that some kind of phonological representation is the input for each morpheme. One challenge is how exceptional phonological properties are to be expressed, since the standard model is that EVAL will select the form that least violates the constraints, thereby normalizing at least some types of aberrant patterns. One possible approach to representing irregularities is to include constraint violations as part of the input representation, to indicate which constraints the input fails to satisfy. (This approach is taken to the limit in Golston 1996, who argues that inputs consist solely of the relevant constraint violations.)

In syntax the "content of the input" question is more puzzling. One possibility is that the input is extremely enriched, containing words, argument structure, and indications of which word is the subject, the object, etc. An even more enriched input would include some degree of syntactic structure assigned to the sequences of words. Inputs might also be impoverished. For instance, the input might include words but not their order or grammatical relations. In this case, a single input will correspond to a variety of outputs with different meanings (e.g. input {dog, man, bites} would correspond to both Dog bites man and Man bites dog. The view taken in both Chapters 5 and 6 is that the input consists of an ordered sequence of words, but not their grammatical structure. The extreme along these lines would be that the input contains no words at all; that words are inserted by GEN or even after EVAL. Under the latter view, EVAL's role is simply to determine which syntactic structures are well–formed, not whether specific instantiations of those types are well–formed.

Third is the issue of faithfulness between the input and the output. In this chapter, for example, we conflated two aspects of faithfulness under FAITHV and FAITHC. These constraints prevented both the *addition* and the *removal* of elements. However, in some languages these two sides of faithfulness may be ranked independently of each other for the same class of elements: DON'T DELETE: "input elements are in the output" vs. DON'T ADD: "output elements are in the input". Furthermore, faithfulness is a relation found not only between input and output, but also between other pairs, such as the **base** and the **reduplicant**, a point illustrated at some length in Chapter 4. The current most prevalent view of faithfulness is **Correspondence Theory**, laid out in McCarthy & Prince (1995) and McCarthy (1995).

Finally, there is the question of whether there is an input at all. Some works have argued that instead of input representations, **morphemes** are best expressed as constraints themselves (Hammond 1995, Russell 1995). As such, they may be ranked with respect to other (nonmorphemic) constraints. An advantage of this approach is the ease with which exceptional behavior is expressed; a disadvantage is that it does not obviously extend to include forms, like sentences, that are larger than the morpheme. The nature of the input is discussed in greater detail in the Afterword .

GEN. GEN's function is to produce a candidate set for every input. There are two aspects of GEN that raise concern. First, in the purely formal model, for every input, an infinite candidate set is generated. Although this does not raise serious problems for formal research, it does hamper efforts to explore psycholinguistic and computational models of language, since neither responds happily to infinite sets. The second problem area is understanding the types of manipulations that GEN can make. It is widely assumed that GEN can only create universally well–formed linguistic objects, that is ones which do not violate any universally inviolable constraints. This assumption requires that we distinguish between universally inviolable constraints and those which are violated, even if only rarely.

CON. The central issues involving the constraint set CON revolve around the question "what are the constraints on the constraints?" Proposals about CON include the idea that certain constraints contain variables which are filled on a language particular basis. (For example, the ALIGN constraints match edges of a pair of elements, where the pair is named in each constraint. See McCarthy & Prince 1993b on Alignment.) Another proposal is that constraints may be conjoined, for instance by logical operators, to make more complex constraints. Yet another challenge, already discussed, is to establish that each proposed constraint is a universal. Chapter 4 raises questions about the universality of each constraint.

EVAL. EVAL evaluates candidates in terms of a particular ranking of constraints, so better understanding of EVAL involves a better understanding of constraint ranking. Typically an analysis will include two or more constraints whose ranking is irrelevant, yet OT assumes that all constraints are ranked with respect to each other. In some work it has been argued that constraints are **tied**, which may allow more than one candidate to be selected as optimal. See Chapter 6 for an example of tied constraints. Another issue involving constraint ranking is "inherent" ranking, where the substantive properties of the constraints themselves determining their ranking with respect to certain other constraint. For example, constraints governing the syllabic positions of elements with different degrees of sonority have been proposed as an instance of inherently ranked constraints (Prince & Smolensky 1993).

The output. The central concern with the output is "what happens next?" In the standard generativist view, the output of one component serves as the input to the next. The metaphor assumes a modular picture of language, where the output of one module serves as the input to others. Is this the best metaphor under OT? For example, the discussion above has focused solely on phonological properties of the string. Phonetic properties might be analyzed through a separate constraint hierarchy, but they might also be analyzed through constraints that intermingle with the phonological constraints. Work by Donca Steriade and her students at the University of California in Los Angeles explores this view.

The modular nature of language. A widely held assumption about linguistic representations and relations is modularity (Fodor 1983). The basic idea of modularity is that the principles responsible for different aspects of an utterance are themselves structured differently. This view is greatly exaggerated under OT, since each constraint can be viewed as an independent entity with its own internal structure.[7] Concerns about modularity arise in another way, too. A logical extension of the OT model for language is that there is a single constraint hierarchy, which internally ranks all constraints, whether syntactic, morphological, phonological, phonetic, or semantic. This possibility predicts interaction between components (**modules**). For example, particular syntactic constraints might be violated in order to satisfy a phonological or morphological constraint, or vice versa. This contrasts sharply with the view of grammar as having a separate and independent syntactic component, phonological component, etc.

Extensions

Within linguistics, some of the most interesting research areas opened up by OT are the interface areas, just mentioned. OT allows the possibility of a single constraint hierarchy, with constraints of all types potentially mingled together. In particular, constraints from different components may be crucially ranked with respect to each other. This possibility provides a new framework for exploring the interfaces between components of the grammar, for example morphology and phonology, syntax and morphology, phonology and phonetics.[8]

There are numerous other domains that may be fruitfully explored using OT, beyond simply the characterization of core language phenomena. Some of these are sketched below.

Poetics. An intriguing domain in which linguistic studies have long played a significant role is the exploration of what is or is not significant to a particular type of poetry. For example, the prototypical relation between poetic meter and spoken rhythm is that strong matches strong and weak matches weak. Studying the matches and mismatches between the strong or weak positions in poetic meter and the strong or weak positions in a nonpoetic rendition of a line of poetry leads to a very precise characterization of the poet's "voice". Recent work by Bruce Hayes and his students at the University of California at Los Angeles suggests that OT offers exciting new insights into the relation between word stress and metrical structure of English folk verse.

Behavior of borrowed words. When words are borrowed from one language to another whose sound patterns are different, the word typically is modified. For example, as noted in our discussion of Hawaiian, that language adapts words borrowed from English, a language which violates Hawaiian syllable constraints, by adding extra vowels: *weleweka* 'velvet'. The expectation under OT is that the borrowing language's constraint hierarchy will evaluate the candidates produced by GEN, taking as input the source language's output of the borrowed word. Exploration of how well this hypothesis works, and where it fails, may lead to significant new insights into what happens when words are borrowed.

---

[7]There are also "families" of constraints, such as the FAITHFULNESS family, for example FAITHV and FAITHC. These constraints all have the same structure and refer to the same type of element. A more precise statement, then, is that each constraint *family* can be viewed as an independent entity with its own internal structure. See especially Chapter 3 for discussion of constraint families.
[8]Works in these areas can be found in the Rutgers Optimality Archive; see the Foreword.

Second language acquisition. When an adult learns a second language, typically the second language is spoken with some degree of "accent". Understanding the nature of that accent is complex, depending on a multitude of variables (such as familiarity with the language) which are difficult to measure. OT provides one guide to identifying patterns we might expect in specific accents, by identifying the constraint rankings of the native language and the second language.

The empirical problem of (first) language acquisition. Under OT, part of acquiring a language is acquiring the critical constraint rankings of that language. Since constraints interact, it is reasonable to assume that evidence for a particular ranking of constraints is not always noticed by the learner, so some constraints are ranked incorrectly, to be reranked when further data is available. This predicts specific stages that a child might go through, each of which would reflect the incorrect dominance of some universal constraint. This prediction is quite different from that of a rule–based model, in which a child might incorrectly learn a language–particular rule, which in itself may have little claim to universality.

The logical problem of language acquisition. This point refers to the challenge of understanding how a child might acquire language under a specific formal model of language. It answers the question of whether a grammar is *in principle* learnable, rather than addressing the issue of how a language is acquired. There is already a small but growing body of work in this area indicating that OT does provide a learnable model of grammars, in particular by Paul Smolensky at Johns Hopkins University and his student Bruce Tesar at the University of Colorado at Boulder, and by Douglas Pulleyblank and his student William Terkel, both at the University of British Columbia.

Language change. Under OT, the formal characterization of language change through time is that constraints are reranked. A prevalent view of diachronic language change is that change occurs when there is imperfect transmission from one generation to the next. Combining these two claims implies that constraints can only be reranked when the evidence for a particular ranking is not very robust. Thus, OT makes clear predictions both about the effects of change and about the type of change that might occur.

Natural language perception. As noted earlier in this section, OT works both from the input to the output and from the output to the input. Under OT it is possible to examine an output and determine the optimal input, something that is not possible under rule–based views of language. For language perception, this is an exciting result, for OT offers a formal theory of language which is able to use outputs to access inputs, crucial to any complete model of language perception.

Natural language production. The standard generative model of phonology, in which an input is manipulated by a series of rules to produce an output, is not readily translated to a model of natural language production. There are two types of challenges, and under OT, both of these problems are solved. First, the formal device of a series of rule applications does not carry over easily into a model of production. Under OT, the input–output relation is mediated in one step, by EVAL, not in multiple steps by a series of rules: this issue, then, does not even arise. Second, the types of rules necessary under such a model include ones which can operate from the end of the word towards the beginning of the word, yet evidence shows that planning for word production (as well as the actual articulation of a word) starts at the beginning of the word, not the end. Under

OT, such apparently directional patterns are determined by inspection of the input and output representations, not as the result of specific operations on representations.

Computational modeling of language. Except for the problem of the infinite candidate set, OT is particularly conducive to computational modeling. Already there are a variety of efforts at developing computational models of specific aspects of OT grammars; OT work in language learnability also relies heavily on computational modeling. This contrasts sharply with other models of phonology, in which each rule of a language expresses an idiosyncratic property of that language, making computational modeling highly idiosyncratic as well.

In closing, I have tried to show that OT not only is gaining wide acceptance among formal linguists, but also that it has extensions into numerous related domains, some of which have already proven to be fruitful. For many, the extensions are as exciting as the successes in formally accounting for core language phenomena.

2
Optimality Theory and Prosody[*]

Michael Hammond

Prosody refers to the organization of sounds into larger phonological units. These include syllables, which are groupings of sounds, and feet, which are groupings of syllables. OT has had its most immediate success in the study of these units.

In this chapter, I review the basic character of the theory in this area and give some arguments for why one would want to treat prosody in terms of OT. I use data primarily from English, for two reasons. First, since you're reading this chapter, I can assume you have some familiarity with English. Second, my own research area includes English, and using English examples will give me an opportunity to present a few novel arguments for an OT treatment of prosody.

1 Unconscious Knowledge

In the domain of phonology, speakers of any language have at least two kinds of unconscious knowledge about their language: they know what the words of the language are and they know what could be a word of the language. For example, in English, we have contrasts between the "words" in (2.1). Words like these elicit the following kinds of responses from a native speaker of English. Word (2.1a) is an occurring word of English. Word (2.1b) could not possibly be a word of English. Finally, word (2.1c) is not an actually occurring word of English, but in principle could be one.

(2.1)   a. *twin*
        b. *tkin*
        c. *trin*

[*]Thanks to Diana Archangeli, Terry Langendoen, Andrea Massar, and Diane Ohala for useful discussion.

You can elicit these judgments in any controlled context.[1] My personal favorite is in an introductory Linguistics course, where these intuitions can be collected en masse. The judgments could also be elicited in a controlled psycholinguistic experiment with all the usual counterbalancing, statistics, etc. These judgments are robust and changes in the methodology don't change the fundamental facts.

Why do linguists want to characterize this as *unconscious* knowledge? It must be unconscious because naive speakers are typically unaware of it. When you walk into a room filled with students or you accost your spouse in the middle of the night and say, "Gee, which sounds better, *trin* or *tkin*?", they're not going to respond, "You know, I've thought about that before, and I've often thought that it might have to do with the onset." Unless they're gifted and a career in linguistics awaits, the more characteristic response is something like: "Huh, I guess *trin* sounds better, but I have no idea why". They have no conscious knowledge that enables them to make the distinction between (2.1b) and (2.1c).

Why is this "knowledge"? One could counter that the reason that *tkin* is an impossible word of English is because it is simply too hard to say. On this view, the unacceptability of *tkin* has nothing to do with knowledge, but everything to do with anatomy. It's easy to show that the unacceptability of *tkin* has nothing to do with anatomy, however. There are other languages where words like *tkin* are completely unremarkable. For example, in Russian, words like *tkin* are common, e.g. *tkan^y* 'fabric', or similar words like *kniga* 'book', *t^yma* 'darkness'. This contrast between English and Russian leads to different predictions. Under the "unconscious knowledge" view, English speakers know one thing and Russian speakers know another. Alternatively, under the "anatomical" view, Russians have anatomical structures that make words like *tkin* easier for them to pronounce. This latter possibility is demonstrably false, however. Native English speakers of Russian ancestry find *tkin* just as unacceptable; and native Russian speakers of English ancestry have no difficulty pronouncing words like *tkan^y*.

> The unacceptability of *tkin* as a word of English is a consequence of unconscious knowledge about English.

In this chapter, I focus on how OT characterizes a native speaker's unconscious knowledge of the syllables and stress patterns in his or her language. Syllables and stress comprise the core of the prosody of a language and are part of the subdiscipline of linguistics known as phonology, which deals with the sound patterns of a language. In the following, I first define syllables and then provide arguments for an OT treatment of them. I then do the same with feet, which are the units of analysis describing stress.

---

[1] The attentive reader will notice that in rapid speech, a word like *tequila* can be pronounced as something like *tkila*. The example *tkin* is only acceptable in rapid speech and only if it has a variant like *tequila* in slower speech, e.g. *takin*.

> Linguistics is a theory that attempts to characterize and account for the acquisition of the unconscious knowledge a speaker has of his or her language.

2 OT and Syllables

Syllables are consonants and vowels grouped into peaks of sonority or intrinsic loudness. We can diagram this as in (2.2), which shows peaks of sonority for the word *sardine* (which has a convenient number of consonants and vowels).

(2.2)   sardine: ⋀⋀
                      [sardin]

It has been claimed that all words are composed of syllables.[2] Following standard usage, I term this claim Syllabic Licensing, and I will assume that there is a constraint corresponding to this claim: LICENSING. In English there is an added requirement, that there be a vowel in every syllable: PEAK (from Chapter 1).

The question arises again: is this anatomy or is this unconscious knowledge? Are Syllabic Licensing and the vowel requirement a simple consequence of anatomy?

It is fairly easy to demonstrate that the PEAK constraint on English syllables must be a consequence of knowledge, rather than physiology. There are many languages where the peaks of syllables need not contain vowels. For example, there are dialects of Berber where there are whole words with no vowels at all: *trglt* 'lock', *txdmt* 'gather wood' (cited in Chapter 1).

The LICENSING constraint is much more difficult to dismiss on the basis of the kinds of facts that we have discussed so far. However, there are arguments that it too is subject to cross–linguistic variation and thus not simply a consequence of anatomy. For example, Bagemihl (1991) cites data from Bella Coola, a Native American language spoken in coastal British Columbia, that suggests that not all sounds in a word are affiliated with syllables. Bagemihl's evidence concerns a process of reduplication (see Chapter 4) which copies the first syllable of a word to mark various categories. A form like *qayt* 'hat' becomes *qaqayti* 'little hat'. However, a form like *qpsta* 'to taste' becomes *qpstata* 'to taste repeatedly', not **qpstaqpsta*. Bagemihl's analysis of this involves claiming that the sequence of consonants at the beginning of *qpsta* — [qps] — is not affiliated with any syllable. If this is correct, Bella Coola stands as an exception to across–the–board LICENSING and an argument that LICENSING cannot be physiological in origin.

> The existence and nature of syllables are not simply a consequence of physiology. They are linguistic constructs.

---

[2]See, for example, Hooper (1972), Kahn (1976), and Itô (1989). See also Chapter 1.

The Factorial Typology

Since syllables are a reflection of unconscious knowledge, it is appropriate to try to model them within linguistic theory, specifically within OT. In this section, I explore the typology of syllabic systems predicted by three of the constraints introduced in Chapter 1 and argue that OT does provide insight into syllabification.

The three constraints are as follows. First, there is the ONSET constraint, which requires all syllables begin with at least one consonant.

(2.3)   ONSET: Syllables begin with a consonant.

Second, there is the preference for syllables not to end with a consonant: NoCODA.

(2.4)   NoCODA: Syllables end with a vowel.

Third, there is FAITHFULNESS, which militates for no changes in the mapping from input to output.

(2.5)   FAITHFULNESS: Pronounce everything as is.

We will see that these constraints provide for a simple characterization of the basic syllabification systems of the world. In addition, these constraints provide for the basis of a simple description of English syllabification.

Let us work through the rankings available here. Notice first ONSET and NoCODA don't directly interact. That is, if they are ranked right next to each other, either ranking of one with respect to the other has the same effect. Hence, there are four possible empirically distinct rankings, as given in (2.6). FAITHFULNESS can be ranked above the other two, as in (2.6a). FAITHFULNESS can be ranked below the other two, as in (2.6b). Finally, FAITHFULNESS is ranked between the other two, as in (2.6c) and (2.6d). Since ONSET and NoCODA do not interact, these are the only possible rankings.

(2.6)   All rankings of {FAITHFULNESS, ONSET, NoCODA}

	rankings	types
a.	FAITHFULNESS » ONSET, NoCODA	(O)V(C)
b.	ONSET, NoCODA » FAITHFULNESS	OV
c.	ONSET » FAITHFULNESS » NoCODA	OV(C)
d.	NoCODA » FAITHFULNESS » ONSET	(O)V

Consider now the right side of (2.6). (O = onset, V = vowel, and C = coda.) These are the kinds of syllables that these rankings predict. The notation is interpreted as follows. If a symbol is present, that means that that structural position is required in a syllable, e.g. all syllables require vowels. If a symbol is absent, that means that that structural position cannot occur in that language, e.g. OV means codas are not allowed. a symbol in parentheses means that that structural position is optional, e.g. (C) means codas are possible but not required (that is, OV(C) is shorthand for OV and OVC in the same language).

Next, let's see how these rankings produce their respective syllable types. The first one (2.6a) is the basic English pattern. FAITHFULNESS is ranked above ONSET and NoCODA, which means that onsets may be missing (*ban* [bæn] vs. *Anne* [æn]) and codas may be present but only in order to satisfy FAITHFULNESS (*no* [no] vs. *note* [not]). The next case (2.6b) holds that ONSET and NOCODA are ranked above FAITHFULNESS. In such a language, the requirement that syllables have onsets and that they not have codas is paramount. A speaker of a language like (2.6b), confronted with *Anne* [æn], would reject such things as words.

The interleaving of FAITHFULNESS with the other two constraints results in "hybrid" systems. If ONSET alone outranks FAITHFULNESS, (2.6c), the onset is required while codas are still possible. If NOCODAalone outranks FAITHFULNESS, codas are forbidden while onsets are optional.

The upshot is that only four types of languages are predicted and, significantly, each of these four types occurs. There is an example of each in (2.7).

(2.7)   **Examples of each type of language**
      a. (O)V(C)   English
      b. OV   Senufo (Guinea)
      c. OV(C)   Yawelmani (California)
      d. (O)V   Hawaiian

Other syllable types are imaginable, but are nonoccurring.

(2.8)   **Impossible languages**

		onset		
		Ø	*required*	*optional*
coda	Ø	V	–	–
	required	VC	OVC	(O)VC
	optional	V(C)	–	–

There is no ranking of the constraints proposed that will give rise to these as languages. The theory does not allow for any of these as language types. That these constraints, freely ranked, can describe all and only the set of possible syllable type systems was first argued by Prince& Smolensky (1993). In general, for any set of freely rankable constraints, OT predicts the possibility of languages exhibiting each possible ranking. This is called the factorial typology.

Granted, the syllable *tokens* in (2.8) occur within languages. English has syllables which are instances of just a vowel and coda, e.g. *Anne* [æn]. But English is not that *type* of language; English is *(O)V(C)*. These tokens are just instances of the general type *(O)V(C)* in English. There is no language which states that *all* syllables must have codas or must be vowel–initial. Such language types simply never occur.

The fact that this nice typological result emerges as a natural consequence is an argument in favor of the general theoretical framework. Placing this analysis in the context of the innateness hypothesis (see Chapter 1), the claim is that children come into the world with these expectations (constraints). Learning a language involves ranking the constraints. Whatever ranking is acquired necessarily produces one of the four possibilities in (2.7). The languages in (2.8) will never exist, for they cannot be learned.[3]

A Rule–Based Alternative

Is there an alternative theoretical approach to compare this one to? One such alternative is to provide a particular procedure or recipe to end up with good syllables. These procedures are called rules.

(2.9)   **Rules:** Syllables are constructed in a step–by–step fashion, gradually
               converting unsyllabified strings into syllabified strings.

The three rules in (2.10) show how English syllabification can be handled in such a framework (V = vowel; C = consonant).[4]

(2.10)   **A rule–based approach**
     a. Vowel Rule: V → [V]
     b. Onset Rule: C[V] → [CV]
     c. Coda Rule: V...]C → V...C]

These rules apply one after the other, in these steps. The first step makes every vowel a syllable. Second, if there is a consonant or string of consonants to the left of that vowel, attach it to the syllable (2.10b) as an onset. Third, if there is a string of consonants left over to the right of the syllable, attach them to the syllable as a coda (2.10c). Some examples are *sardine*, *happy*, and *Joey*.

(2.11)   **English (*sardine*, *happy*, *Joey*)**

	unsyllabified	sardin	hæpi	joi
a.	Vowel Rule	s[a]rd[i]n	h[æ]p[i]	j[o][i]
b.	Onset Rule	[sa]r[di]n	[hæ][pi]	[jo][i]
c.	Coda Rule	[sar][din]	[hæ][pi]	[jo][i]

Notice that these rules are crucially ordered. If the Coda Rule applied before the Onset Rule, we would end up with the wrong results in a word like *sardine*, e.g. *[sard.in]. Words like *card* show that [rd] is a permissible coda in English, yet the Onset Rule has priority in including the [d], as part of the second syllable in *sardine*.

Rules are input–oriented. An illustration of this is found in the application of the Onset Rule to *Joey*: the /i/ does not have a consonant on the left, and so the Onset Rule has no

---

[3]The full typology of syllable systems is, of course, much more complex than (2.6). There are other constraints which must also be interleaved among the three in (2.6) to accommodate systems like Bella Coola and Berber discussed above. The fundamental claim of (2.6) remains nonetheless: there are only four types of systems with respect to permissibility of onsets and codas.
[4]This analysis is parallel to the OT analysis in that it begins with an unsyllabified string of segments and ends with syllables. Other rule–based analyses are possible, but they cannot be compared fairly with the OT analysis provided.

effect. In this way, rules are opportunistic. Each rule has a chance to apply; whether or not a rule actually does apply depends on properties of the input.

> Rules are input oriented. They take an input and tell you what changes to make: whatever results is the output.

A Comparison

We return to the typological facts in order to compare OT with the rule–based model. If one theory captures the attested typology and the other one does not, then this would constitute an argument for that theory over the other.

In the first column of (2.12), we repeat the four types of languages we identified in (2.6), based on the types of syllables found in them. The second column shows which of the three rules of (2.10) must be active to account for each of these types.

(2.12)   **The rule–based typology**

	type	required rules	required success?
a.	OV	vowel, onset	onset, vowel
b.	(O)V(C)	vowel, onset, coda	vowel
c.	(O)V	vowel, onset	vowel
d.	OV(C)	vowel, onset, coda	onset, vowel

The problem for the rule–based theory is summarized in the third column. First, in all these language types, the Vowel Rule must succeed. Second, if onsets are required, as in (2.12a,d), the Onset Rule must also succeed. However, because of the opportunistic nature of rules, their success cannot be guaranteed. If the input lacks a consonant that can serve as the onset for some syllable, rule (2.10b) simply does not apply. In order to capture the typology of syllabification systems, therefore, we must violate the basic idea behind rule systems.

Changing the opportunistic nature of rules has two problems. First, requiring that a rule succeed when it applies is, in effect, imposing an output requirement. As noted above, being output–driven is the earmark of constraints, not of rules.

> Under a rule–based account, the only way to predict the typology is to change rules into constraints.

Second, the rule theory does not easily exclude the nonoccurring language types in (2.8). Consider, for example, *(O)VC*, an impossible syllabic system. This is easily described using rule theory if the Onset Rule is opportunistic and the Coda Rule exhibits "required success". Nothing about the rule analysis prevents the Coda Rule from also partaking of this property with unfortunate empirical consequences. That the typology emerges naturally from the OT analysis and only with significant fudging under the rule–based analysis is an argument in favor of the OT analysis of syllables.

Complications

English is actually more complex than we have presented it above.

I characterized English as a *(O)V(C)* language, but this oversimplifies the situation in two respects. First, as noted above, there are at least two additional constraints which seem to be unviolated in English: PEAK and LICENSING (see Chapter 1). We can add these at the top of the constraint hierarchy as in (2.13) below.[5]

(2.13) PEAK, LICENSING » FAITHFULNESS » ONSET, NOCODA

A second problem is that English allows more than a single consonant in onset or coda position. This can be seen in three contexts. First, the existence of word–initial consonant clusters, e.g. in *twin*, *bring*, *play* shows that onsets can have more than one consonant. Second, the existence of word–final consonant clusters shows that codas can be complex, e.g. in *milk*, *tank*, *carve*. Third, there are word–medial consonant clusters that are at least three consonants long, e.g. in *cylindrical*, *hamster*, *antler*.

To accommodate these facts, we must make two assumptions. First, we must assume that the *COMPLEX constraint from Chapter 1, repeated as (2.14) below, is dominated by FAITHFULNESS.

(2.14) *COMPLEX: Syllables have at most one consonant at an edge.

Second, we must provide some account for why not all sequences of consonants are well–formed. For example, while *tank* is a well–formed English word, *takn* is not. Likewise, while *twin* is well–formed, *tkin* is not. The most obvious generalization governing possible sequences of consonants at the edges of syllables is that they must exhibit an appropriate sonority profile. Onset consonant sequences must increase in sonority and coda consonant sequences must decrease in sonority. This can be formalized as the SONORITY constraint, which also appears to be unviolated in English.

(2.15) SONORITY: Onsets must increase and codas must decrease in sonority.

There are a number of other restrictions on the content of onsets and codas as well. They are of two types. Some govern the degree of rise and fall in sonority. For example, in English while the fall in sonority from [m] to [p] is sufficient for a coda, e.g. *limp*, the rise in sonority from [p] back to [m] is insufficient for an onset. On the other hand, the fall from [r] to [p] is sufficient for a coda, e.g. *harp*, and the rise from [p] to [r] is sufficient for an onset, e.g. *pry*. A second class of restrictions limits the similarity between two consonants in the same onset or coda. For example, sequences involving the lips like [pw] or [bw] are generally avoided in onsets, with only a few borrowed exceptions, e.g. *bwana* and *pueblo*. We'll not treat these more specific restrictions here.

---

[5]A more detailed analysis shows that these two constraints are outranked by others to accommodate syllabic [l, r] and consonant clusters with [s] respectively. We leave these complications aside here.

Notice also that character of word–initial and word–final clusters only follows on the assumption that all segments satisfy LICENSING in English. Hence, the absence of *tkin* as a possible word of English follows from two assumptions. First, all segments must satisfy LICENSING. Second, all syllables must satisfy SONORITY. The existence of words like *tkan^y* in Russian is a consequence of lower ranking of one or the other of these two constraints (presumably LICENSING).

To summarize so far, English is an example of an *(O)V(C)* language with FAITHFULNESS ranked above ONSET and NOCODA. English exhibits complex onsets and codas as well, entailing that *COMPLEX is low ranked. These clusters satisfy sonority sequencing which entails that LICENSING and SONORITY are high ranked. Finally, every syllable must have a vowel, which can be captured by ranking PEAK highly. A complete hierarchy consistent with these conclusions is given as (2.16).

(2.16)   PEAK, LICENSING, SONORITY » FAITH'NESS » ONSET, NOCODA, *COMPLEX

Confirmation

In this section, I review two arguments that support the theory of syllabification provided above and the specific analysis of English

First, in Chapter 1 a general argument for the syllable as a unit of cognitive organization was provided: it allows us to characterize the possible words of any language more simply. This argument also applies to English.

> All polysyllabic words can be analyzed as sequences of well–formed syllables.

A more specific argument for the existence of syllable is based on where a profane or obscene expression (henceforth expletive, as in 'expletive deleted') can occur within a word of English, for emphasis, or to express surprise or disgust. For example, by this rule, the word *Minnesota* can be realized as *Minne–fuckin–sota*. The appropriate context can be seen with an imagined vignette: a close friend tells you he is moving from Arizona to Minnesota for the sun. You might respond as follows (assuming an appropriate level of informality holds between you and your confused interlocutor): "You're going to Minnesota? Minne–fuckin–sota?" Speakers of British English can do this as well using a comparable expletive: *Minne–bloody–sota*.[6]

There are interesting limits on where you can put the expletive though. Consider the word *sardine*. In *sardine*, there are six sounds which we can represent by the letters s-a-r-d-i-n, and five places (indicated by the dashes) where in principle the expletive could be positioned internal to the word, yet only one of these places are acceptable. I've marked the ungrammatical forms with an asterisk (*).

The expletive must occur in a position such that the material on each side is well syllabified in terms of the restrictions on English. If more than one well syllabified option is available, then the expletive goes where the split portions are *best* syllabified.

---

[6]According to Fussell (1975), expletives first were used within words in English by soldiers in World War I.

(2.17)　　Argument #2: Expletive infixation
    a.　*s–fuckin–ardin
    b.　*sa–fuckin–rdin
    c.　sar–fuckin–din
    d.　*sard–fuckin–in
    e.　*sardi–fuckin–n

Case (2.17d) is the most interesting; [sard] and [in] are both well–formed syllables. However, they are not as well–formed as the spans in (2.17c): [sar] and [din]. This can be clearly seen in the following tableau where these two syllabifications are compared.

(2.18)

/sardin/	ONSET	NoCODA
☞　sar.din		**
sard.in	*!	***

Hence *sar–fuckin–dine* is the preferred infixed form.

This is confirmed with cases having only a single intervocalic consonant, e.g. *propane*. A form like **prop–fuckin–ane* requires less optimal syllabification, hence *pro–fuckin–pane* is the preferred infixed form.

(2.19)

/propen/	ONSET	NoCODA
☞　pro.pen		*
prop.en	*!	**

Thus the facts of expletive infixation confirm the analysis of English syllables provided above. First, determining the position of the expletive requires reference to the syllable, confirming that a description of syllables is required. Second, the position of the expletive is based on the best syllabification of the word in question, confirming that an optimality–based analysis is superior.

Notice, however, that we have not told the whole story about expletive infixation. While a word like *propane* can undergo it, a word like *happy* cannot: **ha–fuckin–ppy* and **happ–fuckin–y* are both ungrammatical, even though *happy* is made up of two syllables. This problem is dealt with in the next section.

3 OT and Stress

In this section, I first say what stress is and then provide some reasons for thinking that it is part of people's unconscious knowledge of their language. Then I provide an account of several stress–based phenomena in English and argue that they are best treated in terms of constraints, rather than rules.

Stress

What is stress? Syllables are organized into patterns of alternating prominence, loudness, length, or pitch. In English, each pattern of alternation (termed a foot) contains a stressed syllable on the left and at most one stressless syllable on the right. This particular kind of foot is termed a trochee; other languages have different kinds of feet with different internal arrangements of stressed and stressless syllables. The most common two patterns are trochees, as in English; and iambs, a stressless–stressed sequence of syllables.

The prominence pattern of the word *Minnesota* can be diagrammed as in (2.20). There are two prominent syllables, [mi] and [so], each followed by a less prominent syllable, [nə] and [tə] (where [ə] "schwa" is a neutral vowel). Thus the word can be analyzed as made up of the feet as in [Mìnne][sóta], where I use brackets to mark feet and accents to mark prominent vowels.[7]

(2.20)
Minnesota

The same question arises as for the syllable: Are there really groups of syllables, or are these alternations just an acoustic effect? The arguments are parallel.

First, the character of words is best described in terms of feet. That is, just as sounds must be licensed by syllables, syllables must be licensed by feet. The situation is a little more complex than in the case of syllables, however. Syllables are not all necessarily parsed into feet. Rather, words are parsed into a series of one or more feet with at most one stressless syllable intervening between any foot and the edge of the domain or between any two feet. We will term this partial foot licensing.

Let's consider some examples to see how this works. A word like *hát* bears a stress. A single stressed syllable is a legal trochee; hence *hát* satisfies partial foot licensing: [hát]. A word like *háppy* has stress on the first syllable. A stressed syllable followed by a stressless syllable is a legal trochee; hence *háppy* satisfies partial foot licensing: [háppy]. A word like *abóut* has stress on the second syllable. That second syllable can be treated as a foot. The word exhibits partial foot licensing because the foot is separated from the beginning of the word by only a single stressless syllable: a[bóut]. A word like *América* has stress on the second syllable. The second and third syllables can be grouped together in a foot. The word satisfies partial foot licensing because only a single unfooted syllable separates the foot from the beginning of the word and from the end of the word: A[méri]ca.

Let's consider some more complex cases. As I have already shown, *Mìnnesóta* has stresses on the first and third syllables, and can be analyzed as made up of two feet each with a stressed and a stressless syllable: [Mìnne][sóta]. A word like *Wìnnepesáukee* has stresses on the first and fourth syllables. Two feet can be built with only a single stressless syllable intervening: [Wìnne]pe[sáukee].

Partial foot licensing excludes some possibilities as well. Consider first a hypothetical form like *America* with stress *only* on the third syllable. This could only be footed by leaving the first two syllables stray, violating partial foot licensing and predicting that

---

[7]I use an acute accent (´) to mark the most prominent vowel, and a grave accent (`) to mark the next most prominent vowels. This distinction in prominence is not analyzed here.

such a form is ungrammatical: *Ame[ríca]. Consider also a hypothetical form *América* with stress only on the first syllable. Here again, the most complete footing leaves two adjacent syllables stray, predicting that such a form would be ungrammatical: *[Áme]rica. Finally, imagine a hypothetical form *Winnetepesáukee* where two syllables must be left stray in the middle of the word: *[Wìnne]tepe[sáukee]. Again, such a form is correctly predicted to be ungrammatical.[8]

This array of facts can be treated with the constraints in (2.21)–(2.23). First, words must be stressed.[9] Second, feet in English are trochaic; i.e. stressed on the first syllable. Finally, words cannot have two unfooted syllables in a row.[10] With these constraints, we can express the partial foot licensing restriction. Consider first a monosyllabic word like *hát*. ROOTING requires that such a word bear stress, as shown in (2.24)

(2.21)   ROOTING: Words must be stressed.

(2.22)   TROCHEE: Feet are trochaic.

(2.23)   PARSE–SYLLABLE: Two unfooted syllables cannot be adjacent.

(2.24)

/hæt/	ROOT	TROCH	PARSESYLL
☞ [hæt]			
hæt	*!		

With a disyllabic word, a number of possible stress patterns emerge as equally well–formed. To see this, we consider a hypothetical form *pata*, as in (2.25). There are two observations to make about this example. First, notice that the system predicts multiple outputs for a single input, yet speakers of English know that for the overwhelming majority of specific words, only one stress pattern is possible; hence *abóut* can never be pronounced *ábout*. This fact can be treated with additional constraints that cause particular syllables to attract stress.

(2.25)

/pata/	ROOT	TROCH	PARSESYLL	examples
pata	*!		*	*
☞ [páta]				háppy?
☞ [pá]ta				háppy?
☞ pa[tá]				abóut
☞ [pá][tá]				rébòund
[patá]		*!		*

---

[8]Partial foot licensing does not appear to be required with certain affixes. Hence the word *cóniferless* 'without conifers' is acceptable, even though two stressless syllables separate the foot from the right edge of the word: [cóni]ferless.

[9]The name of this constraint comes from Hammond (1984). In OT, this constraint is typically referred to as LxWD≈PrWD.

[10]This is not the best statement of this constraint. However, a better statement of it would require a treatment of stress attraction. Certain syllables seem to attract stress either exceptionally or because of the segments in them. This is an extremely complex issue and we leave it aside.

Two classes of constraints are needed. One class attracts stress to syllables closed by a coda or containing certain vowels in certain positions. For example the second syllable of *agénda* must be stressed; a word with that form could never be stressed *ágenda*. Second, there are certain words that simply require stresses in certain positions. The proper treatment of these restrictions has been debated since the earliest days of generative phonology and we will not treat them here. The critical observation is that these additional restrictions do not alter the consequences of the constraint set we have presented above. All English words can be described with these constraints.

A second observation to make about (2.25) is that two of the patterns are indistinguishable on the surface: [páta] and [pá]ta. Since these are apparently indistinct, let us posit a constraint to eliminate this redundancy. The basic observation is that a monosyllabic foot before a stressless syllable is equivalent to a disyllabic foot in the same context. Therefore we posit the **BINARITY** constraint in (2.26), which assigns a violation in the case of [páta]. If BINARITY is added to the constraint set, tableau (2.25) expands to (2.27)..

(2.26) **BINARITY**: A monosyllabic foot cannot occur before an unfooted syllable.

(2.27)

/pata/	ROOT	TROCH	PRSSYL	BIN	*examples*
pata	*!		*		*
☞ [páta]					háppy
[pá]ta				*!	*
☞ pa[tá]					abóut
☞ [pá][tá]					rébòund
[patá]		*!			*

As a final example, let us consider the range of possibilities for a trisyllabic span. Again, we find that all the occurring patterns are correctly generated, as shown in (2.28).

(2.28)

/pataka/	ROOT	TROCH	PRSSYL	BIN	*examples*
pataka	*!		*		*
pa[tá]ka				*!	*
pata[ká]			*!		*
☞ pa[tá][ká]					eléctròn
☞ pa[táka]					vanílla
pa[taká]		*!			*
[pá]taka			*!	*	*
[pá]ta[ká]				*!	*
☞ [pá][tá][ká]					chìmpànzée
☞ [pá][táka]					bàndána
[pá][taká]		*!			*
☞ [páta]ka					Cánada
[patá]ka		*!			*
☞ [páta][ká]					Tènnessée
[patá][ká]		*!			*

A second argument that feet in English are not a simple acoustic effect is that expletive infixation can only take place between feet, as intially argued by McCarthy (1982); see also Hammond (1991). Hence *Min–fuckin–nesota* is unacceptable; *Minne–fuckin–sota* is O.K.; but *Minneso–fuckin–ta* is not acceptable. This is diagrammed in (2.29) below.

(2.29)   **Argument #2: Expletive infixation**
    a. *Mi–fuckin–nnesota*
    b. Minne–fuckin–sota
    c. *Minneso–fuckin–ta*

This correlation between the boundary required by the description of where the stresses go and the places where you can place the expletive holds for words of all stress configurations. Consider an example like *Wìnnepesáukee*. Here a stressless syllable intervenes between the two feet leaving two positions for expletive infixation. Expletive infixation in both locations is acceptable.

(2.30)   [Wìnne]pe[sáukee]
    Wìnne–fuckin–pesáukee
    Wìnnepe–fuckin–sáukee

In fact, the proposal appears to make the correct predictions about even more complex examples like *Tìmbùktú* or *hàmamèlidánthemum*, with three feet each.

(2.31)   **a.**   [hàma][mèli][dánthe]mum
           hàma–fuckin–mèlidánthemum
           hàmamèli–fuckin–dánthemum
    b.   [Tìm][bùk][tú]
           Tìm–fuckin–bùktú
           Tìmbùk–fuckin–tú

The intuitions of naive subjects are in fact quite robust, even with unusual words, and provide striking confirmation for the proposal that words are organized into feet.

Another Argument: Syncope

In this section, I provide an argument that a constraint–based analysis offers the best treatment of stress as well. This argument is based on syncope in fast speech, a process whereby certain vowels can disappear in particular positions in English words; see Zwicky (1972), Pérez (1991) for discussion of some of these facts. In (2.32), a stressless vowel can syncopate in fast speech.

The kinds of consonants that occur on each side are important, the character of the vowel matters, the frequency or familiarity of the word plays a role, and the rate of speech. These are not the conditions I'm interested in here. I therefore only provide examples where the consonants on each side are appropriate, and will assume that the examples are sufficiently familiar and uttered at an appropriate rate. Making these assumptions will allow us to focus on the stress conditions.

(2.32) **Fast speech syncope**

		slow	fast
a.	*At the beginning of words:*	paráde	práde
		Torónto	Trónto
		Marína	Mrína
		Canádian	Cnádian
b.	*Before a stressless syllable:*	ópera	ópra
		géneral	génral
		chócolate	chóclate
		réspiratòry	réspratòry
		glòrificátion	glòrficátion
c.	*After a stressless syllable and*	réspiratòry	réspirtòry
	before a stressed syllable:	glòrificátion	glòrifcátion

What are the stress conditions on syncope? The vowel that exhibits syncope must be stressless. The vowel can be at the beginning of the word, as in (2.32a). It can also occur in the middle of a word before a string of one or more stressless syllables, as in (2.32b). Thus, *opera* almost always becomes [opra], *general* becomes [*genral*], and *chocolate* becomes [*choclate*]. In longer words, more options are possible. For example, *respiratory* can surface as [respirtory] or [respratory]. Some comparisons are given in (2.33).

(2.33) óp(e)ra   òperátic
       gén(e)ral   gènerálity
       glórify   glòr(i)f(i)cátion
       réspiràte   resp(i)r(a)tòry

Here, parentheses mark cases where the vowel syncopates readily. Thus the middle vowel of *opera* is easily elided, but in *operatic*, the corresponding vowel must remain: *[opratic] is egregiously bad. In *general*, the middle vowel can syncopate, but not in *generality*. Notice especially that syncope is not a simple function of how long the word is. We can try to state these stress conditions in terms of a rule, (2.34).

(2.34) **A rule–based account:** Syncopate a stressless vowel if
    a. it is word–initial; or
    b. it immediately follows the only stress of the word and is not in the last syllable of the word; or
    c. it is one of two stressless syllables intervening between two stressed syllables.

Consider (2.34) carefully. Notice that all three environments must be stipulated in (2.34). The problem for the rule–based treatment is that there are three separate clauses. If we can restate the process in such a way that there is only one clause, this would be an improvement. (This is, of course, simply Occam's razor.)

Using OT, we can do just that. Basically, the syncope rule in (2.34) misses the generalization that vowels syncopate only when an optimal (=disyllabic) foot would result.

The environments above can only be reduced to a single environment when the *output* is considered.

For example, a word like *parade* has stress on the second syllable. This second syllable is a foot, and the first syllable is unfooted: pa[ráde]. That homeless syllable can syncopate and it improves footing, in the sense that more of the word is footed: Licensing is enhanced, e.g. pa[ráde] -> [práde]. In constraint–based terms, the pressure to foot syllables exceeds the pressure to pronounce some of them.

What about the second case, *opera*? *Opera* starts off with three syllables, so one syllable is left unfooted if no vowel is deleted: [ópe]ra. When the medial vowel syncopates, we end up with *opra*, which is now a lawful foot: [ópra]. In the third case of where this happens, *respiratory*, there is a stress on [tor] and a stress on [res]. There are five syllables; each stressed syllable is grouped with the following syllable, e.g. [tory] and [respi]. Notice that if we delete one of the vowels in the middle, the word can be completely footed, either as [respir][tory] or [respra][tory].

The critical point is that in all three cases — *parade*, *opera*, and *respiratory* — deleting a vowel results in improved "Foot Licensing" (*FOOTLESS below). If, on the other hand, we state a rule that deletes vowels in particular contexts in terms of input, this generalization is not apparent. This improved coverage can be seen when we consider the foot structures (square brackets) for the forms in (2.33) above. Syncopated vowels are not counted in foot structure.

(2.35)   [óp(e)ra]   [òpe][rátic]
         [gén(e)ral]   [gène][ráli]ty
         [glóri][fȳ]   [glòr(i)fi][cátion], [glòrif(i)][cátion]
         [réspi][ràte]   [résp(i)ra][tòry], [réspir(a)][tòry]

Let's formalize this in terms of a set of strictly ranked constraints to see that we obtain the right results. The central constraint in this system is *FOOTLESS (2.36b), which prefers that all syllables be footed. *FOOTLESS is thus a stronger version of PARSE–SYLLABLE (2.23) and exactly parallel to LICENSING.

To allow some vowels to be deleted, we must decompose FAITHFULNESS a little. First, let us assume that the pronunciation of stressed and stressless vowels in fast speech is governed by separate subconstraints of the FAITHFULNESS family. We call these two constraints FAITH(ó) and FAITH(ŏ). Both of these constraints militate for faithfulness between fast speech and normal speech. Specifically, they impose a correspondence between vowels in fast speech with their respective normal speech analogs.[11] Thus FAITH(ó) requires that vowels that are stressed in normal speech be pronounced in fast speech. However, FAITH(ŏ) requires that vowels that are stressless in normal speech be pronounced in fast speech.

The fact that stressless vowels can syncopate to avoid violations of *FOOTLESS, but not stressed vowels, can be captured by ranking FAITH(ó) above *FOOTLESS and FAITH(ŏ) below *FOOTLESS. The desired ranking is given in (2.37).

---

[11]For discussion of correspondence, see McCarthy & Prince (1995), and Chapter 4.

(2.36)   **A constraint–based account:**
   a. FAITH(ó): pronounce stressed vowels.
   b. *FOOTLESS: no unfooted syllables.
   c. FAITH(ŏ): pronounce unstressed vowels.

(2.37)   FAITH(ó) » *FOOTLESS » FAITH(ŏ)

We have to assume that there are also other constraints that control syllable structure and familiarity of the words that outrank the constraints in (2.36) and that prevent whole-sale deletion of any stressless vowel. The basic result, though, is that a stressless vowel can be syncopated just in case this results in less unfooted material in the word, i.e. fewer violations of *FOOTLESS.

The analysis as given doesn't quite succeed yet. To see this, let us first consider how this would work for some representative cases in (2.32). Examples like *glorify* with no syncope work straightforwardly. Cases like *respiratory* are also straightforward. Notice how the constraint system proposed results in a tie, correctly predicting two possible pro-nunciations in this speech rate (Hammond 1994).

(2.38)

/glorify/	FAITH(ó)	*FOOTLESS	FAITH(ŏ)
gl(ó)ri[fȳ]	*!	*	
[glór](i)[fȳ]			*!
[glórif](ȳ)	*!		
☞ [glóri][fȳ]			

(2.39)

/respiratory/	FAITH(ó)	*FOOTLESS	FAITH(ŏ)
r(é)spira[tòry]	*!	**	
☞ [résp(i)ra][tòry]			*
☞ [réspir](a)[tòry]			*
[réspi]rat(ò)ry	*!	**	
[réspi]ra[tòr](y)		*!	*

The problem arises with words like *general*. Consider the following tableau, where two candidates emerge as optimal, but only one is acceptable. (The "❽" identifies the incor-rectly selected candidate.)

(2.40)

/general/	FAITH(ó)	*FOOTLESS	FAITH(ŏ)
g(é)neral	*!	**	
☞ [gén(e)ral]			*
❽☞ [géner(a)l]			*
[géne]ral		*!	

If we want to avoid this consequence, we must augment the system with an additional constraint that requires that final syllables be pronounced regardless of their stress. This constraint, call it FAITH(FINAL), must be ranked above *FOOTLESS, as in (2.41).

(2.41)   FAITH(ó), FAITH(FINAL) » *FOOTLESS » FAITH(ŏ)

The correct output now results for *general*, as shown in tableau (2.42).

(2.42)

/general/	F(ó)	F(F)	*F'LESS	F(ŏ)
g(é)neral	*!		**	
☞ [gén(e)ral]				*
[géner(a)l]		*!		*
[géne]ral			*!	

FAITH(FINAL) is likely a reflection of a general lengthening of final syllables regardless of their stress (Fowler 1977). Presumably, this length inhibits syncope.

The OT analysis of syncope thus has the following character. First, there is a difference between stressed and stressless vowels such that stressed vowels resist syncope. This is encoded by separating out FAITH(ó) from FAITH(ŏ) and ranking the former above the latter. Second, final syllables resist syncope even when stressless and this is formalized by separating out FAITH(FINAL) and also ranking it highly.

The central pressure for syncope is formalized as a constraint *FOOTLESS. In conjunction with the constraints previously posited, this has as a natural consequence that only the syllables in the positions indicated in (2.32) can syncopate.

The OT analysis proposed describes the data in a fashion superior to the rule–based alternative proposed in (2.34). This is because the constraints proposed are a simpler and more direct statement of the generalization underlying the process than the statement in (2.34). The rule–based account of (2.34) could only stipulate particular configurations in which vowels would delete. As we have seen, however, these environments can be unified when we consider the *output* of syncope, rather than the input.

4 Interaction of Stress and Syllabification

In this section, we provide additional confirmation for the analyses of syllabification and stress in this chapter, by considering how they interact.

Some Observations about Stress and Syllabification

As discussed in Chapter 3, sounds exhibit different pronunciations in different contexts. In this section, we consider one such case in English where the pronunciation of a set of sounds is a function of prosodic structure. The particular facts to be discussed here have been discussed extensively in much previous work on English. See, for example, Kahn (1976), Selkirk (1982), and Borowsky (1986).

At the beginning of words, /t/ is pronounced with aspiration, a puff of air coinciding with a delay in the onset of vocal cord vibration for the following voiced sound. This aspiration does not occur when /t/ follows /s/ and does not occur at the ends of words.[12]

(2.43) a. top   [tʰap]
        b. stop [stap]
        c. post [post]

This regularity is easily treated if we say that aspiration occurs at the beginning of syllables, which of course provides yet another argument for syllables.

However, the facts are somewhat more complex word–medially. If the following syllable is stressed, /t/ is aspirated. If the following syllable is unstressed, /t/ is unaspirated. (In this unaspirated context, /t/ can even surface as a voiced flap [ɾ] in American English, which is indicated in the transcriptions below.)

(2.44) a. stressed–stressless:
            batter [bǽɾər]
        b. stressless–stressless:
            vanity [vǽnəri]
        c. stressed–stressed:
            hotel [hòtʰél]
        d. stressless–stressed:
            attack [ətʰǽk]

It has been proposed that the distribution of aspiration medially can be captured if we suppose that the edges of syllables care about stress. Specifically, we can maintain the form of the aspiration generalization — that /t/ is aspirated syllable–initially — if we propose that an intervocalic consonant is affiliated to the left before a stressless syllable and to the right with a stressed syllable. These syllabifications are indicated with dots in (2.45), and below.

One might have thought a foot–based account of aspiration was viable: /t/ is aspirated at the left edge of a foot. Examples like *tomórrow* with word–initial aspiration before a stressless syllable show that this won't work. See Kiparsky (1977) and Hammond (1982) for discussion.

(2.45)

medial	ó‿ó	V.CV	[hò.tʰél]	hotel
	ó‿ŏ	VC.V	[bǽɾ.ər]	batter
	ŏ‿ó	V.CV	[ə.tʰǽk]	attack
	ŏ‿ŏ	VC.V	[vǽnər.i]	vanity
initial	word[‿ó	.CV	[.tʰǽki]	tacky
	word[‿ŏ	.CV	[.tʰəmáro]	tomato

---

[12]Virtually the same regularity applies to /p/ and /k/, but the acoustic distinction is clearer with /t/ and there are small complications with the different sounds, so we'll confine our attention to /t/.

An OT Analysis of the Interaction between Stress and Syllabification

Suppose we posit a constraint called NoOnset, which outranks Onset and NoCoda, and which forces an intervocalic consonant to affiliate with the syllable to its left if the following vowel is stressless.

(2.46) NoOnset: Stressless medial syllables are onsetless.

Let's consider now how NoOnset gets the right results in the hypothetical form *pata*. We assume that the constraints forcing stress in particular positions outrank the constraints affiliating the intervocalic consonants in question. The legal stress configurations are thus evaluated separately.

(2.47)

/pata/	NoOnset	Onset	NoCoda
[pá.ta]	*!		
☞ [pát.a]		*	*
☞ pa.[tá]			
pat.[á]		*!	*
☞ [pá].[tá]			
[pát].[á]		*!	*

This proposal about the syllabic affiliation of intervocalic consonants is supported by other data. First, there is additional linguistic evidence from other sources that reinforce this conclusion (Kahn 1976; Borowsky 1986; Myers 1987), e.g. the pronunciation of intervocalic [h] and postconsonantal [u]/[yu].

Consider the case of [h]. This sound can occur at the beginnings of words, but not at the ends. Thus we have words like *happy*, *Henry*, *Homer*, etc. This sound never occurs at the ends of words.[13] Between vowels, it has precisely the distribution one would expect given NoOnset. Before a stressed vowel it is pronounced, as in *vehícular*, but before a stressless vowel it is not pronounced, as in *véhicle*. Hence, the pronunciation of /h/ confirms the proposal here about the syllabic affiliation of intervocalic consonants.

Independent Confirmation from Speech Perception

In this section, we consider a rather different argument for the OT treatment of intervocalic syllabification. This argument comes from the interaction of stress and syllabification in speech perception.[14] So far, we have considered prosody from a strictly phonological perspective. Our goal has been to understand how syllables and feet — and an OT characterization of them — figure in a treatment of a native speaker's unconscious

---

[13]There are a number of words spelled with *h* on the right, but this is purely orthographic. The *h* in these positions is either not pronounced at all, e.g. *Sarah*, or is pronounced as something else, e.g. *catch*, *wish*, *myth*.

[14]This argument is developed in full detail in Hammond & Dupoux (to appear).

knowledge of their language. In this section, we consider how syllables and feet might figure in a treatment of how people perceive speech.

There is strong evidence that the syllable plays a role in online speech perception. One psycholinguistic task, fragment monitoring, has been extensively explored. In this methodology, subjects are asked to listen for strings of sounds at the beginning of words. If they detect the string at the beginning of a word, they are to press a button as quickly as possible. Reaction times are collected and analyzed. For example, a subject might be instructed to monitor for, listen for, the string [kan]. If the subject hears the word *conic* [kánɪk] or *contact* [kántækt], they would press their button, but not if they hear the word *tonic* or *nocturne*. If they hear the word *table*, they would not press the button.

By using this technique and carefully controlling the kind of materials subjects are presented, a clever argument for the role of the syllable in speech perception in French has been constructed (Cutler et al. 1983, 1986). Subjects listen for CV or CVC strings like [pa] or [pal]. The relevant words contain *both* strings, e.g. *palace* 'palace' or *palmier* 'palm tree'. The difference between the two word classes is where the edge of the first syllable is. In words of the *palace* type, the edge of the first syllable is after [pa], [pa.las]; in words of the *palmier* type, the edge of the first syllable is after [pal], [pal.mye].

Subjects are significantly faster when they are asked to find a string that ends at the edge of a syllable, than if not. Results from one such experiment (Cutler et al. 1983) are graphed in (2.48), in which reaction time in milliseconds to the two types of carrier words is plotted against the two types of target sequences. There is a statistically significant interaction between target type and word type. Subjects are significantly slower in detecting CVC targets in words like *palace* than they are at detecting CV targets in them. This is reversed in *palmier*–type words. In words like *palmier*, subjects are significantly slower at detecting CV targets than CVC targets.

This effect can be accounted for if we assume that the syllable plays a role in online speech perception. Assume that at some level of processing subjects break a word up into syllables. Assume too that this same level is accessed by the fragment monitoring task. Subjects are faster at detecting CV targets in *palace*–type words because the target matches the parsing units exactly. They are slower with CVC targets because the target does not match the parsing unit; to detect [pal] in *palace*, a subject needs to inspect more than the first unit *palace* is parsed into: [pa]. The opposite situation obtains with *palmier*–type words. Subjects are faster when detecting CVC targets because they exactly match the first unit such words are parsed into, e.g. [pal]. To detect a CV target, the subject must further analyze this chunk, hence predicting the observed slower reaction time. These facts thus establish the relevance of syllables for speech perception.

Consider now the results of analogous experiments with analogous materials with English–speaking subjects (Cutler et al. 1983; 1986; and Bradley et al. 1993). It has been shown that there is no interaction when CV and CVC targets are sought in words of the type *balance* vs. *balcony*. Results from Bradley et al. (1993) are graphed in (2.49). Again, reaction times are plotted against target types for different word types. There is no significant interaction between the two factors. Thus there is no evidence from experiments of this sort that the syllable plays any role in speech perception for English–speaking subjects.

(2.48)   **French subjects, French words**

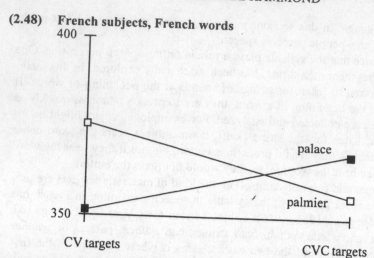

However, this conclusion is premised on the assumption that the syllabification of words like *balance* and *balcony* is parallel to their French analogs. That is, the absence of a syllabic effect in English only holds if one can assume that the first syllable of *balance* is CV, just like the first syllable of French *palace*. Similarly, this conclusion depends on the first syllable of *balcony* being CVC just like the first syllable of *palmier*.

As we argued in the previous section, there is compelling linguistic evidence that the [l] in a word like *balance* should be syllabified to the left. Moreover, off–line judgments of syllabic affiliation also show that stress is a significant factor in the affiliation of inter-vocalic consonants (Treiman & Danis 1988, Treiman & Zukowsky 1990).

Thus we might expect a significant effect with other English materials. Hammond & Dupoux (to appear) report on a fragment monitoring task with English–speaking subjects where CV and CVC targets were crossed with words of varying stress, e.g. *clímàx* vs. *clímate*. The [m] in *clímàx* should affiliate to the right and the [m] in *clímate* should af-filiate to the left. Specifically, we would expect CV targets to be detected more rapidly in words of the form *climax*, while CVC targets should be detected more rapidly in words of the form *climate*. However, no significant interaction is found.

Let's assume that stress does play a role, but an *occluding* role rather than a deriva-tional role.[15] This, in fact, is the position suggested by Cutler et al. and is the position to be explored here. Cutler et al. propose that there is no significant effect of syllable struc-ture in English speech perception because the syllable structure of English is unclear. It is unclear because normal syllabification by ONSET (as in French) is at odds with stress–based syllabification by NOONSET.

---

[15]There are several other possible conclusions one could draw. First, it could be that while stress is a significant factor, there are other stronger factors that prevent the stress factor from reaching significance. This hypothesis is explored in depth in Hammond (1993), where the role of vowel quality is pursued, and we will have nothing more to say about it here. It is an empirical question whether other factors overpower stress in fragment monitoring in English. Second, it could be that the syllable that the grammar makes use of is simply different from the syllable that speech perception makes use of. This would have to be the position of last resort.

(2.49)   English subjects, English words

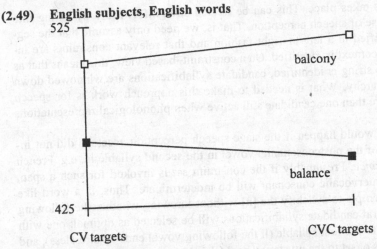

This latter position seems quite reasonable, but is impossible to formalize in a rule–based framework. I argue here that if we are to adopt this account of the difference between English and French, then we are committed to an OT–based treatment of English syllabification. The argument goes like this.

If we assume that there is a rule enforcing stress–based syllabification, then it must apply at some point in the derivation. If the representations provided by linguistic theory are the same as those referred to by speech perception, then it must be the case that syllables before or after the stress–based syllabification rule are referred to. The problem is that in either case, the wrong predictions are made. If there are syllables before stress–based syllabification and those are the syllables referred to by speech perception (2.50), then presumably they are analogous to French syllables and we would get an effect with materials analogous to French materials. As we saw above, this is not the case. Hence, syllables before stress–based syllabification are not referred to by speech perception.

(2.50)   Prestress syllabification: *ba.lance* vs. *bal.co.ny,* but *cli.max = cli.mate*

On the other hand, if stress–based syllable are referred to by speech perception, then we expect an effect with the *climax/climate* contrast (2.51). No such effect surfaces. Hence, speech perception does not access a post stress–based syllabification syllable.

(2.51)   Poststress syllabification: *bal.ance = bal.co.ny,* but *cli.max* vs. *clim.ate*

Since neither stage provides a satisfactory syllable that will account for the facts of English, one of two things follows. First, linguistic representations may be irrelevant. As noted already above, parsimony would urge us to make this a last resort. Second, linguistic *derivations* may be mistaken. In other words, a workable solution may be found if we assume that syllable structure is a consequence of ranked constraints, rather than ordered rules of syllabification.

We can account for the absence of an effect with the English–speaking subjects if it turns out that there is no single representation available at the point the identification of syllable–sized chunks takes place. This can be treated straightforwardly by taking into account the time course of speech perception. That is, we need only assume that the segmental string is identified in a left–to–right fashion and that relevant constraints are invoked as the relevant context is identified. On a constraint–based view, this means that as more and more of the string is identified, candidate syllabifications are winnowed down by the constraint hierarchy. What is needed to make this approach work is for speech perception to find more than one candidate still active when phonological representations are accessed.

Consider now what would happen if the stage speech perception accessed did not include the stress value of the postconsonantal vowel in the second syllable in, e.g. French *palace* or English *balance*. Presumably if the constraint set is invoked for such a span, the affiliation of the intervocalic consonant will be indeterminate. Thus, in a word like *póny* [póni], if one attempts to syllabify the [n] without knowing whether the following vowel is stressed, several candidate syllabifications will be selected as optimal: one with [n] affiliated as a coda to the first syllable (if the following vowel ends up stressless), and another where the [n] is onset to the mystery vowel (should it turn up stressed).

To see this, let us consider the analysis of English in a little more detail. In the previous section, we proposed that English syllable structure was governed partially by the constraints ONSET and NOCODA. The facts of aspiration told us that there is some constraint that outranks these and moves an intervocalic consonant to the left if the following vowel is stressless: NOONSET. Consider now what happens if this constraint set is invoked on a partially processed string where the stress value of the following vowel is as yet undetermined.

(2.52)

/pónV/	NOONSET	ONSET/NOCODA
☞ pó.nV		
☞ pón.V		*

Since the stress of the final vowel cannot be determined, violations of NOONSET cannot be assessed. Therefore violations of lower ranked constraints also cannot be assessed. (This follows because any lower ranked violation might be overruled by a violation of NOONSET; therefore lower ranked violations cannot be assessed until the higher ranked violations are clear.) Unassessable constraints are marked with shading. Thus both syllabifications indicated in (2.52) must be acceptable at this point.

Such an intermediate position is completely consonant with the psycholinguistic results. The absence of an effect in English is because presumably at the stage where fragment monitoring takes place, the candidate set hasn't resolved to a single candidate. This is not the case in French of course. In French, NOONSET does not play a role. (It is either absent or lower ranked than ONSET.) Hence if this is the stage that is accessed by speech perception, only a single candidate will be available.

(2.53)

/pónV/	ONSET/NOCODA
☞ pó.nV	
pon.V	*!

Such a position is impossible on the rule–based view. At all stages of the derivation, relevant rules have either applied or not.

One might think that this account is problematic because it hinges on access occurring at a specific point in the processing of the word. However, there is good reason to take this position. First, in this experimental method, subjects are explicitly told to press the button as soon as they identify the target string. Second, part of the task involves identifying the relevant consonant. Since formant transitions are a strong cue for the identity of consonants, it is not surprising that subjects should make their decision having processed at least some of the information contained in the following vowel. Finally, we know from a variety of sources that stress values are assessed nonlocally (Lehiste 1970). That is, whether a vowel is stressed or not is something that can only be judged by comparing that vowel to neighboring vowels. Hence it is not unreasonable to suppose that the determination of the stress value of the following vowel might take longer. All of these considerations support the position that accessing the syllabic structure when the following vowel is only partially identified is a reasonable supposition.

> An OT treatment of prosody can be extended easily to accommodate results outside of phonology proper.

This thus constitutes an additional argument in favor of an OT treatment of prosody. If we make the assumption that the representations that phonology manipulates are the same ones manipulated by speech perception, it then follows that we must manipulate those representations in terms of OT.[16]

5 Conclusion

Let's summarize the arguments we have given from prosody in favor of OT.

First, in the syllable domain, the OT analysis of syllables extended in an obvious way to provide the typology of syllables in the word's languages. A second argument coming from stress was that the location of syncope is more readily stated in terms of the structures that result instead of the structures before syncope. Finally, a third argument came from the joint effects of syllables and stress in speech perception, where it was argued that only an OT–based analysis accounts for the absence of a syllabic effect. This latter argument is of course the most speculative as it involves looking outside of the traditional areas of phonological concern.

There are many other domains where prosody plays a role. There are OT analyses of some of these, but not all.

Child speech production. Gerken (1994) argues that children omit certain syllables when they speak and the pattern of omission is best treated in terms of feet. Massar (1996) has argued that these data are best treated in terms of OT. Ohala (1996) argues

---

[16]The OT–based syllabification algorithm discussed here is implemented computationally. The program can be run via the World Wwide Web: **http://aruba.ccit.arizona.edu/~hammond**.

that the simplification of consonant clusters by children is also best treated in terms of OT.

Lexical access. Cutler & Norris (1988) argue that syllables play a role in the process whereby people extract words from the speech stream. Meador (1996) investigates the role of stress and vowel quality in this domain.

Poetic meter. Syllables and stress play an obvious role in regulated verse. Hayes & McEachern (1996) have provided an OT analysis of folk meters. Other OT treatments of metrical phenomena include Fitzgerald (1995) and Elzinga (1996).

Language games. Pig Latin also make clear reference to stress and syllable structure. Bagemihl (1988) is a wonderful survey of the kinds of systems that occur. Hammond (1990) and Davis & Hammond (1995) discuss some language games of English in rule–based terms.

Finally, morphology interacts with stress and syllable structure in a very dramatic way. There is an extensive OT literature already on this subject; see Chapter 4 for further discussion.

It would not be unreasonable to expect that applying OT to these other domains could result in significant explanatory advances as well.

3
Optimality Theory and Features[*]

Douglas Pulleyblank

This introduction to featural patterns in Optimality Theory begins by identifying the fundamental function of speech sounds, their role in providing access to meaning. The simplest and most effective way of identifying meanings is to have a unique and invariant phonological address for every distinct meaning. Hearing a particular set of sounds from some language would invariably direct a speaker of that language to a particular meaning. But such a simple mapping of sounds onto meanings is not what we observe in natural language. The complexity that is actually found in sound systems results from the need to reconcile the demands of conflicting constraints on the organization of speech sounds. This chapter focuses on identifying some of these constraints, and showing how Optimality Theory can resolve conflicts between them.

1 Decomposing Sounds into Features

It is a standard observation that sentences are composed of words, that words are composed of smaller meaningful units referred to as **morphemes** (*talent, talent+ed, un+talent+ed; judge, judge+ment, judge+ment+al; art, art+ist, art+ist+ic*), and that morphemes themselves are composed of strings of sounds; see also Chapter 4. Compare the word *a* with a single sound, *at* with two, *pat* with three, *spat* with four, and so on.

[*]Thanks to Darin Howe and Myles Leitch for comments on a draft of this chapter, as well as to participants in a Holland Institute of Generative Linguistics seminar at the University of Amsterdam for their questions and comments on sections presented there. In addition, a special acknowledgment is due to Diana Archangeli and Terry Langendoen for extremely detailed comments both on the chapter's form and on its substance. The work was supported by Social Sciences & Humanities Research Council of Canada, grant #410–94–0035, and completed under the auspices of a grant from the Nederlandse organisatie voor wetenschappelijk onderzoek.

However, the decomposition does not stop there. Speech sounds are themselves composed of smaller elements called **features**. Compare, for example, the consonants found in the following words: *pin, zip, bin, man, fin, vet, dad, kin, sit, gas*. Although each of these words is composed of three sounds or **segments**, the features that make up the various sounds crosscut the different consonants. For example, the consonants [p,b,f,v,m] are all **labial**, articulated with the lower lip; [t,d,s,z,n] are all **coronal** (or **alveolar**), articulated by raising the tongue tip or blade to the alveolar ridge, the bony ridge immediately behind the upper teeth; [k,g] are **dorsal** (or **velar**), produced by raising the tongue body (or dorsum) to the soft palate (also called the velum). The classification of such consonants according to distinctions based on these and other **places of articulation** plays a role in numerous phonological patterns.

(3.1)   **Classification based on place of articulation**

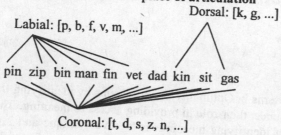

The same consonants form different sets if classified according to different featural dimensions. For example, the consonants [b,v,m,d,z,n,g] are all **voiced**, produced with vibrating vocal cords; in contrast, the consonants [p,f,t,s,k] are all **voiceless**, produced with vocal cords that are sufficiently spread apart to prevent any vibration.

(3.2)   **Classification based on voicing**

In a similar fashion, these consonants can be classified according to nasality: **oral** sounds are produced with all airflow through the oral cavity [p,b,f,v,t,d,s,z,k,g]; **nasal** sounds are produced with air flow through the nasal cavity [m,n].

Finally, the speech sounds in these words can be classified as to whether the airflow in the oral cavity is completely obstructed and then released (**stops**) [p,b,m,t,d,n,k,g], or whether airflow is continuous and with friction (**fricatives**) [f,v,s,z].

Each distinct speech sound combines a different set of such featural specifications. Hence [f] is labial, voiceless, oral, and a fricative; [n] is coronal, voiced, nasal, and a stop; [g] is dorsal, voiced, oral, and a stop; and so on.

(3.3)   **Classification based on nasality**

(3.4)   **Classification based on continuancy**

> A speech sound results from a combination of subsegmental features, each feature defined by particular articulatory and auditory properties.

2 Identity Hypothesis: INPUT = OUTPUT

Consider now a fundamental problem of the sound–meaning relationship that is crucial to language. Sounds themselves do not have meaning; particular *sequences* of sounds have meanings. For example, the three individual sounds of *pin* and *vet* have no meanings on their own, but when combined together are identified with particular meanings. (See also Chapter 4.) But how are such meanings accessed? Meanings are associated with particular sets of speech sounds in a set of **lexical entries**. These lexical entries essentially form a mental dictionary: they relate sounds to meanings, provide syntactic information (the word is a verb; takes a direct object; etc.), provide information about related words, and so on. The lexical entry contains some sequence of sounds (the **input**) that is related to the sounds occurring in some fully formed utterance (the **output**). The input is paired with a meaning; the output occurs in an actual speech event. Given such a picture, it is possible to postulate an extremely simple mapping between the two levels of representation: the input is identical to the output. According to this hypothesis, once a *listener* identifies some outpu string in an utterance, the string directly constitutes a pointer for locating a meaning. For a *speaker*, the input to output relation involved in going from meaning to sound is similarly direct.

The hypothesis that the input is identical to the output is not correct, however, in many cases. For example, consider the prefix *in-*, meaning 'not': *elegant* vs. *in+elegant*, *efficient* vs. *in+efficient*, *appropriate* vs. *in+appropriate*, etc. Although these examples are consistent with the hypothesis that Input *in-* = Output *in-*, a consideration of additional examples shows that this is not always the case. For example, while the consonant of the affix is pronounced [n] before coronal consonants such as [t,d] (*tolerable* vs. *in+tolerable, definite* vs. *in+definite*), the same input [n] corresponds to the sound [m] in cases like *perfect* vs. *im+perfect, balanced* vs. *im+balanced*, in which the consonant immediately following the prefix is labial. A simple generalization governs all these cases: the nasal has the same place of articulation as the following consonant. That is, it is coronal when the following consonant is coronal, labial when the following consonant is labial, and so on.

INPUT:	The speech sounds paired with a meaning in the mental lexicon
OUTPUT:	The speech sounds as they occur in a complete utterance
HYPOTHESIS:	INPUT = OUTPUT

(3.5)   **The INPUT and OUTPUT are not always identical**

INPUT: /in/

OUTPUT: [in...]   [im...]

There is a simple explanation for the deviation from input/output identity. Identity would require a sequence of consonants with differing places of articulation: *in+perfect* where the n+p sequence contains a coronal nasal [n] followed by a labial stop [p]. By abandoning identity, the output form is able to contain a sequence of two labial consonants, [m] followed by [p]: *im+perfect*. As such, the articulators are in a single state throughout both consonants, an instance of articulatory inertia. Cases such as these exhibit a basic tension between the pressure for input/output identity on the one hand, and the pressure to produce a sequence of consonants at the same place of articulation on the other.

Let us consider briefly the manner in which input and output forms are formally related within a grammar. In a **derivational theory**, cases where an input and output are not identical are accounted for by applying a series of rules: the first rule in the series applies to the input form, the second rule applies to the intermediate form derived by the application of the first rule, the third rule applies to the form derived by the second rule, and so on. Once all rules have applied, the resulting form is the output. In the case in (3.5), the output form with [m] would be derived by applying a rule causing the nasal to acquire the same place of articulation as the following consonant. See also Chapters 1 and 4 on the properties of a derivational theory.

Such a theory produces a rather curious expectation. It is generally assumed that a derivational grammar with fewer rules is simpler than a comparable grammar with more rules. For example, a grammar achieving the output forms in (3.5) with one rule would be considered simpler, and therefore more highly valued, than a grammar that accomplished the same result with two, or three, or seventeen rules. That is, all else being equal, the fewer the rules, the better the analysis. But pursuing this logic to its extreme would mean that the simplest grammar would be one where there are no rules, where all inputs are identical to all outputs. In other words, why deviate from identity at all? Isn't the simplest phonology one that isn't? While interpreting fewer rules as simpler might at first seem desirable, there is an immediate and apparent problem: none of the anticipated simple grammars without phonological rules have ever been found. Why should complexity be an apparently unavoidable property of sound systems?

DERIVATIONAL THEORY

LEXICON:	INPUT FORM
RULE 1:	INTERMEDIATE FORM$_1$
RULE 2:	INTERMEDIATE FORM$_2$
:	:
LAST RULE:	FINAL FORM = OUTPUT

Optimality Theory presents a solution to this problem (Prince & Smolensky 1993, McCarthy & Prince 1993a). The core of the theory is conflict resolution. Although a certain class of constraints, the **faithfulness constraints**, tries to impose identity on the relation between input and output, faithful relations will in many instances cause direct conflict with other constraints.[1]

(3.6)   Faithfulness constraints

MAX: Every segment/feature of the input has an identical correspondent in the output

DEP: Every segment/feature of the output has an identical correspondent in the input.

As is demonstrated below, there is no *simplest* grammar: respecting the faithfulness conditions of MAX and DEP leads to violations of other constraints; respecting other constraints leads to violations of faithfulness. Individual grammars resolve this conflict by ranking the relevant constraints in various ways.

---

[1]Correctly determining the formal properties of the faithfulness constraints is an important issue within Optimality Theory. This paper does not address such issues, but assumes a version of faithfulness modeled on the proposals of **correspondence theory** made in McCarthy and Prince (1995). For the introductory purposes of this chapter, it is not necessary to distinguish between MAX and DEP in tableaux, hence both are referred to simply as FAITH. Note that no reference is made to the IDENT (demanding the featural identity between correspondents) family proposed in McCarthy and Prince (1995); throughout this paper, the role that McCarthy and Prince attributed to IDENT is attributed directly to MAX and DEP, by assuming that MAX and DEP directly govern features, a proposal made in Myers (1995).

> Spoken language cannot obey all the constraints on phonological well–formedness.
>
> Individual languages resolve cases of conflict by assigning more importance to some constraints than to others.

3 Deviation from Identity: Syntagmatic Constraints

Consider again the case of the English prefix *in–* which appears in some instances as *in–* and in other instances as *im–*. With this prefix, faithfulness comes into conflict with pressure to have the cluster of two adjacent consonants share a single value for place of articulation. We can describe this pressure by means of a class of **syntagmatic** constraints, so–called because they impose restrictions on possible sequences of sounds. The four featural dimensions discussed in Section 1 give rise to four specific syntagmatic constraints which we call IDENTICAL CLUSTER CONSTRAINTS.

(3.7)   IDENTICAL CLUSTER CONSTRAINTS:
   VOICING: A sequence of consonants must be identical in *voicing*.
   PLACE: A sequence of consonants must be identical in *place of articulation*.
   CONTINUANCY: A sequence of consonants must be identical in *continuancy*.
   NASALITY: A sequence of consonants must be identical in *nasality*.

The function of the IDENTICAL CLUSTER CONSTRAINTS is to impose articulatory inertia — changes in the positioning of articulators are disallowed. A single type of voicing, a single place of articulation, a single type of occlusion, and a single value for oral vs. nasal airflow is required throughout an entire cluster of consonants.

> CONFLICTING WITH IDENTITY
>
> **Syntagmatic constraints** require particular featural properties of sequences of segments.
>
> For example, the IDENTICAL CLUSTER CONSTRAINTS require identity of the featural properties of sequences of consonants.

We consider below how the first three of these four constraints operate in sequences consisting of a nasal consonant followed by an oral consonant, specifically, an oral **obstruent**, either an oral stop {p,b,t,d,...} or an oral fricative {f,v,s,z,...}. A nasal–obstruent sequence as in (3.8) is exactly the type of sequence already observed in the English examples like *in+definite* and *im+perfect*.

The features found in such clusters are given in (3.8). Such clusters by definition have a nasal first consonant and an oral second consonant, hence differ necessarily in their values for the nasal/oral feature. Normally, the nasal itself is a stop (the oral cavity is completely obstructed) while the obstruent may be either a stop or a fricative. In terms of voicing, the nasal is normally voiced, while the obstruent again exhibits variation, occurring either as voiced or voiceless. Finally, the nasal may appear with any place of articulation (signified in (3.8) by "Place *x*", *x* a variable ranging over place of articulation), as may the following obstruent (*y* another variable for place values). In sum, the nasal–obstruent cluster differs in nasality and may or may not differ along the dimensions of continuancy, voicing, and place of articulation.

(3.8) **Features in nasal–obstruent clusters:**
mb, mp, mv, mf, md, mt, mz, ms, mg, mk
nb, np, nv, nf, nd, nt, nz, ns, ng, nk, ...

NASAL	OBSTRUENT
Nasal	Oral
Stop	Stop or Continuant
Voiced	Voiced or Voiceless
Place *x*	Place *y*

In the following subsections, we examine nasal–obstruent clusters in several languages of West Africa and Mexico to see the various ways in which the faithfulness constraints interact with the constraints on identical clusters.

Anything Goes: Faithfulness above Identical Cluster Constraints

Consider first a case where faithfulness constraints outrank all of the syntagmatic constraints on nasal–obstruent clusters in (3.7). Maddieson (1983) provides a large number of examples of nasal–obstruent clusters in words in Bura, a Chadic language of Nigeria, including those in (3.9). (The symbols [š,ž] indicate palatoalveolar fricatives, as in the English words *[sh]ip* and *plea[s]ure*; the symbol [ə] represents a schwa, the central vowel which occurs as the first, unstressed vowel in the English words *p[o]lice*, *p[a]rade*, and *m[e]ridian*. The accents on Bura words indicate **tones**, the use of pitch distinctions to distinguish between different meanings. For further discussion of tone, see Sections 6–8.)

(3.9) **Bura: nasal–obstruent clusters**

Cluster	Word	Gloss
[mp]	mpà	'fight'
[mb]	mbà	'burn'
[mt]	mtà	'death'
[md]	mdâ	'person'
[ms]	msə́kâ	'maternal uncle'
[mš]	mší	'corpse'
[mž]	mžá	'be enough'

As we have already seen, a nasal–obstruent cluster violates the constraint on cluster nasality; in addition, cases like [mt, md, ms, mš, mž] violate cluster place; [mp, mt, ms, mš]

DOUGLAS PULLEYBLANK

violate cluster voicing; and [ms, mš, mž] violate cluster continuancy.[2] Each of these sequences violates anywhere from one to four of the IDENTICAL CLUSTER CONSTRAINTS in (3.7). The following tableau gives an example in which all four of the cluster conditions are violated in order to respect the faithfulness constraints for those features. Note the following abbreviations: NAS = nasal, CONT = continuancy, VOI = voicing, PL = place of articulation.

(3.10)   **Bura: high ranking of faithfulness**

/mši/	FAITHFULNESS CONSTRAINTS				IDENTICAL CLUSTER CONSTRAINTS			
	NAS	CONT	VOI	PL	NAS	CONT	VOI	PL
☞ mši					*	*	*	*
ñši				*!	*	*	*	
mži			*!		*	*		*
mči		*!			*		*	*
ññí	*!	*!	*!	*!				

The optimal candidate is *mši* (the first candidate in (3.10)): being identical to the input, it fully satisfies all faithfulness constraints. The candidate *ñši* (the second candidate in (3.10)) is better with respect to the cluster condition on place of articulation, but is ruled out because it violates faithfulness with respect to place; *mži* (the third candidate in (3.10)) satisfies the cluster condition on voicing, but in so doing violates faithfulness on the same feature; *mči* (the fourth candidate in (3.10)) satisfies the condition on continuancy by converting the obstruent into a stop, but again, such satisfaction comes at an unacceptable price, namely a violation of faithfulness for continuancy. Finally, *ññí* (the fifth candidate in (3.10)) is maximally good in terms of the cluster conditions; by having completely identical consonants, all cluster conditions are satisfied. With respect to the same features, however, all faithfulness conditions are violated, rendering it unacceptable.

> In some languages, faithfulness to the input is more important than having clusters with shared feature values.

Returning briefly to the issue of simplicity, note that the input–output relations in Bura are as would be expected in a derivational grammar without rules. One might expect, therefore, that such relations would be the commonest cross–linguistically. In fact, however, the Bura pattern is quite rare. More common are patterns where at least one of the cluster conditions ranks higher than the relevant faithfulness condition. Some representative cases are considered below.

---

[2]Bura data are simplified here somewhat. For example, the nasal in clusters involving a voiceless consonant [mp], [mt], etc. is voiceless at the beginning of a phrase but voiced in the middle of a phrase.

Homorganicity: Enforcing Agreement in Place Features

With the English prefix *in–*, we have a case in which the nasal consonant of the prefix agrees in the output with the labial feature of the initial consonant of an adjective like *perfect*: Input: *in+perfect* ↔ Output: *im+perfect*. Such cases exhibit **homorganicity**, a situation where a sequence of segments shares a single set of specifications for place of articulation.

Homorganicity is robustly illustrated by Yoruba, a Niger–Congo language of Nigeria (Ward 1952, Pulleyblank 1995). Yoruba differentiates six places of articulation:

- labial, e.g. [b, m];
- labiodental, e.g. [f, ɱ];
- alveolar (coronal), e.g. [t, d, s, n];
- alveopalatal, e.g. [ɟ, ɲ];
- velar (dorsal), e.g. [k, g, ŋ];
- labial–velar (labial + dorsal simultaneously), e.g. [kp, gb, ŋm].

However, when a nasal occurs immediately before a consonant, the nasal must have the same place of articulation as the following consonant. This can be seen clearly in verbs marked for progressive aspect, which is indicated by the attachment of a nasal prefix. The initial consonant of each Yoruba verb root corresponds to the place of articulation indicated in the first column, as does the nasal prefix in each corresponding progressive form. (A tilde (˜) over a segment indicates nasalization, as in the French word *bon* [bɔ̃] 'good'. As in Bura, the accent marks indicate tones; these have no bearing on the place of articulation of either nasals or obstruents.)

(3.11)   **Yoruba: nasal–obstruent clusters**

Place of articulation	Verb root	Progressive form	Gloss
Labial	bá	ḿ+bá	'overtaking'
Labiodental	fɔ́	ɱ́+fɔ́	'breaking'
Alveolar	tà	ń+tà	'selling'
Alveolar	dũ̀	ń+dũ̀	'paining'
Alveolar	sũ̀	ń+sũ̀	'sleeping'
Alveopalatal	ɟó	ɲ́+ɟó	'dancing'
Velar	kɔ	ŋ́+kɔ	'writing'
Velar	gũ̀	ŋ́+gũ̀	'climbing'
Labial–velar	kpa	ŋḿ+kpa	'killing'
Labial–velar	gbɔ́	ŋḿ+gbɔ́	'hearing'

Maintaining input/output identity with respect to place of articulation (FAITH[PLACE]) is in direct conflict with the requirement that a cluster exhibit a single place of articulation (IDENTICAL CLUSTER CONSTRAINT [PLACE]). Ranking the cluster condition above faithfulness results in this subpart of the Yoruba grammar.

(3.12)   IDENTICAL CLUSTER CONSTRAINT [PLACE] » FAITH[PLACE]

> Some languages require that consonant sequences share a single place of articulation even if such sharing produces a discrepancy between the input and the output.

Before providing a tableau illustrating this analysis, one additional issue must be addressed. Consider an input in which a coronal nasal precedes a dorsal stop, as in /...n+k.../.[3] The fully faithful output would maintain the $n+k$ sequence at the cost of violating IDENTICAL CLUSTER CONSTRAINT [PLACE]. To prevent a violation of the cluster constraint, there are two possible options. The coronal nasal could be replaced by a dorsal nasal ([...ŋ+k...]) giving two dorsals, or the dorsal stop could be replaced by a coronal stop ([...n+t...]) giving two coronals. Both options are successful with respect to the cluster constraint, but languages only seem to exhibit the pattern in which the nasal changes. Phrased in derivational terms, the problem is why the place of the obstruent (stop or fricative) spreads to the nasal, and not the other way round.

(3.13)   Retention/loss of place specifications in nasal–obstruent clusters

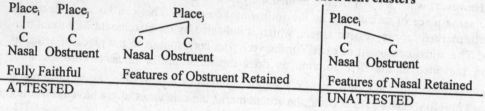

Place$_i$ Place$_j$	Place$_j$	Place$_i$
C C	C C	C C
Nasal Obstruent	Nasal Obstruent	Nasal Obstruent
Fully Faithful	Features of Obstruent Retained	Features of Nasal Retained
ATTESTED		UNATTESTED

Within Optimality Theory, we can say that the faithfulness constraint that enforces identity for place features in obstruents is *universally* ranked above the corresponding faithfulness constraint for nasals (see Pater to appear). That is, OT proposes two types of rankings, (i) those that vary from language to language, and (ii) those that are invariant across languages. Prince& Smolensky (1993) have called invariant rankings between two constraints harmonic rankings. Thus the universal ranking we have just observed is a harmonic ranking. It is one which is motivated phonetically since place distinctions are more perceptually salient on obstruents than on nasals.

(3.14)   Harmonic ranking of faithfulness for place of articulation
FAITH[PLACE] OBSTRUENTS » FAITH[PLACE] NASALS

Given the ranking in (3.14), the correct Yoruba output form, ŋ́kɔ, is unambiguously selected. To maintain separate places of articulation violates IDENTICAL CLUSTER CONSTRAINT for place, *ńkɔ. Losing the place features of either the obstruent or the nasal violates FAITHFULNESS for place, but losing the nasal's place feature, ŋ́kɔ, is better than losing the obstruent's place feature, *ńtɔ.

---

[3]For Yoruba, the assumption that the nasal is coronal /n/ is merely based on markedness, coronal being the preferred place of articulation for consonants cross–linguistically. Given the high ranking of IDENTICAL CLUSTER CONSTRAINT [PLACE], it makes no difference what the place of the progressive nasal is in the input.

(3.15)   Yoruba: ranking of place agreement over faithfulness

/ń–kɔ/	ICC[PLACE]	FAITH[PLACE]OBSTRUENTS	FAITH[PLACE]NASALS
ńkɔ	*!		*
☞ ŋkɔ			
ńtɔ		*!	

Optimality Theory postulates two types of rankings:

VARIABLE RANKINGS	*Different rankings define different languages* LANGUAGE 1: A » B LANGUAGE 2: B » A
HARMONIC (INVARIANT) RANKINGS	*Other constraints may intervene, but the ranking is universal* LANGUAGE 1: C » D LANGUAGE 2: C » X » D

Enforcing Agreement in Voicing

Just as a nasal–obstruent cluster may be required to agree in place of articulation, so may such a cluster be required to agree in voicing. For example, Zoque, a Mixe–Zoquean language of southern Mexico, requires that a stop following a nasal be voiced (Wonderly 1951, 1952, Kenstowicz & Kisseberth 1979, Sagey 1986, Padgett 1994). (The symbols [č,ǰ] represent the sounds of the initial consonants of English *cheap* and *jeep* respectively; [ʌ] is the vowel in the English word *cup*.)

(3.16)   Zoque: voicing of stops

INPUT	OUTPUT	
/min–pa/	[minba]	'he comes'
/min–tam/	[mindamʌ]	'come! (PLURAL)'
/pʌn–čʌki/	[pʌnǰʌki]	'figure of a man'
/pʌn–kʌsi/	[pʌngʌsi]	'on a man'

Like the Yoruba cases just examined, these forms require that a member of the IDENTICAL CLUSTER CONSTRAINT family be ranked above faithfulness. The difference is that the feature concerned is voicing rather than place of articulation. Whereas it is possible for the input in a nasal–obstruent cluster to violate the IDENTICAL CLUSTER CONSTRAINT for voicing by having a voiceless stop follow a voiced nasal, it is necessary that all outputs respect the condition, as in (3.17). To ensure the correct output, it is necessary that the IDENTICAL CLUSTER CONSTRAINT [VOICE] be ranked above FAITH[VOICE]: No matter what the input values for voicing, the output values must obey the cluster condition. This effect is shown in the tableau in (3.18).

(3.17)   **Voicing values in a nasal–stop sequence**

	Nasal	Stop	ICC[VOICE]
INPUT	VOICED [n,...]	VOICELESS [p,t,č,k,...]	*
OUTPUT	VOICED [n,...]	VOICED [b,d, ǰ,g,...]	OK

(3.18)   **Zoque: ranking of voicing agreement over faithfulness**

/min–pa/	ICC[VOICE]	FAITH[VOICE]
minpa	*!	
☞ minba		*

> Some languages require that consonant sequences share a single value for voicing even if such sharing produces a discrepancy between the input and the output.

Note that a third possible candidate would be a form where the nasal undergoes devoicing, rather than the obstruent becoming voiced. This would result in an ungrammatical form like *miṇpa, where the open dot under the *n* indicates voicelessness. Candidates containing voiceless nasals are eliminated in Zoque by invoking a **paradigmatic** feature constraint which forces the cooccurrence of nasality with voicing (see Itô, Mester, & Padgett 1995, Pater to appear). Such constraints are discussed in Section 4.

Combination of Effects

The syntagmatic effects of one feature may interact in various ways with the effects of additional features. In Zoque, we have seen that the consonants in a nasal–obstruent cluster must agree in voicing, but they need not agree in place of articulation: *minba* 'he comes' and *pᴧngᴧsi* 'on a man' in (3.16). The grammar of Zoque, therefore, must rank the IDENTICAL CLUSTER CONSTRAINT for place below faithfulness for place, while the cluster constraint for voicing has the opposite relation to its faithfulness counterpart. Since the morphological distinction between a stem and an affix becomes important in these examples (the need for the distinction will be discussed shortly), stems are indicated in relevant examples by the inclusion of square brackets.

(3.19)   **Zoque: voicing agreement but not place agreement in stems**

[min]Stem+*pa*	FAITH [PLACE] STEM	ICC [PLACE]	ICC [VOICE]	FAITH [VOICE]
[min]pa		*	*!	
[mim]pa	*!		*	
☞ [min]ba		*		*
[mim]ba	*!			*

By ranking FAITH [PLACE] STEM above IDENTICAL CLUSTER CONSTRAINT [PLACE], only those candidates are possible where underlying place specifications are preserved. Hence this ranking preserves the candidates *minpa* and *minba*, but eliminates *mimpa* and *mimba* (candidates that would have been optimal in a language like Yoruba). Comparing the candidates that survive this initial culling, *minba* is preferable to *minpa* because it respects IDENTICAL CLUSTER CONSTRAINT [VOICE] .

Let us return now to the **stem** qualification of place faithfulness. While stems must retain any place specifications that they possess in their inputs, such faithfulness is not enforced within larger domains that include affixes (see McCarthy & Prince to appear, Pater to appear for further discussion from a somewhat different perspective). Consider the behavior of the first person singular prefix in (3.20).

(3.20) **Zoque: some affixes exhibit both voicing and place identity**

INPUT	OUTPUT	
n+[pama]	m[bama]	'my clothing'
n+[tatah]	n[datah]	'my father'
n+[čoʔngoya]	ñ[ĵoʔngoya]	'my rabbit'
n+[kayu]	ŋ[gayu]	'my horse'

In contrast with the data in (3.16), these forms exhibit nasal–obstruent clusters showing agreement both in voicing and in place of articulation. Both the nasal and the obstruent are voiced, and they are both labial or both coronal, and so on. (Note that all output sequences in (3.21) agree in their voicing values because of the effect of ICC[VOICE].)

(3.21) **Zoque: place of articulation values in a nasal–stop sequence**

	WITHIN STEM						OUTSIDE STEM					
INPUT	np	nt	nk	mt	mk	...	np	nt	nk	mt	mk	...
	↕	↕	↕	↕	↕		↕	↕	↕	↕	↕	
OUTPUT	nb	nd	ng	md	mg		mb	nd	ŋg	nd	ŋg	
	• faithful to input for place						• respects ICC[PLACE]					
	• so may violate ICC[PLACE]						• so may violate faithfulness for place					

Such distinctions between root or stem domains and larger domains are not uncommon. Mohanan (1993) suggests that constraints tend to hold more strongly of smaller domains than of larger domains. In the Zoque case under discussion, we must assume that being faithful to the underlying values for place is more important in the stem domain than it is generally. The stem–restricted constraint must be ranked above IDENTICAL CLUSTER CONSTRAINT [PLACE], while the unrestricted constraint must be ranked below.

(3.22) FAITH [PLACE] STEM » ICC [PLACE] » FAITH [PLACE]

The effects of this ranking can be seen in (3.23), where the various candidates exhibit differences in both voicing and place (PL = Place, VOI = Voice).

(3.23)　Zoque: faithfulness is more stringently required of stems than of affixes

n+[pama]	FAITH[PL]STEM	ICC[PL]	ICC[VOI]	FAITH[VOI]	FAITH[PL]
n[pama]		*!	*!		
m[pama]			*!		
n[bama]		*!			*
☞ m[bama]				*	
n[tama]	*!		*	*	*
n[dama]	*!			*	*

The form *npama is ruled out because it violates the cluster constraints on both place and voice: n is coronal and voiced while p is labial and voiceless. The forms *mpama and *nbama are similarly excluded because they each violate one of the cluster constraints: *mpama violates the constraint on voicing while *nbama violates the constraint on place. The optimal form is mbama, which satisfies the cluster constraints on both place and voice.

With regard to the stem/nonstem distinction, note that the input min+pa has an output minba (3.19), where both n and b are faithful to their underlying place value. In contrast, the input n+pama has the output mbama (3.23), where m is unfaithful with respect to place in order to be homorganic to the following labial consonant. In both cases, the input contains a sequence /...n+p.../. The outputs differ, however. In neither case is it possible to change the place value of the stop. This is due to the importance of being faithful to the place values of a stop. The behavior of the nasal varies, however. When the nasal is part of a stem constituent (as in [min]+pa), its place values must be retained; when it is part of an affix (as in n+[pama]), the cluster constraint on place overrides faithfulness.

> A constraint may have a stronger effect in a small domain than in a large domain. For example, a constraint may hold of the stem domain, but not hold of the larger domain including prefixes and suffixes.

Syntagmatically Induced Deletion

In addition to cases where the sequence of segments induces some effect on the particular features of a segment that appear in the output (voicing, place, etc.), certain types of sequences are simply impossible in some languages — and the only way to respect such impossibility is the deletion of a segment. Consider for example the result in Zoque of attaching the first person singular marker to stems beginning with a fricative. As already seen in (3.20), the first person singular prefix is a nasal. In (3.24), the same nasal prefix occurs with stems beginning in a fricative, but the nasal is completely absent from the output.

(3.24)   Zoque: deletion of nasal prefix in nasal–fricative sequences

INPUT	OUTPUT	
n+[sʌk]	[sʌk]	'my beans'
n+[šapun]	[šapun]	'my soap'

Recall that fricatives are produced with continuous airflow through the oral cavity, that is, they are *continuants*. Nasals, on the other hand, have a complete blockage of air through the oral cavity; they are *stops*. The result is that when a nasal immediately precedes a fricative, the resulting cluster has conflicting values for continuancy, thereby violating the IDENTICAL CLUSTER CONSTRAINT [CONTINUANCY]. Different languages of the world have been observed by Padgett (1994) to provide three basic resolutions to this problem: (i) the nasal can become *continuant* (Kpelle, Spanish), (ii) the fricative can become a *stop* (Venda, Zulu), (iii) the nasal can be deleted (Lithuanian, Zoque).[4] The Zoque case of deletion is considered here.

Were a nasal–fricative sequence to appear in the output, it would present a cluster problem: the nasal is a stop while the fricative is continuant. By ranking the cluster constraints prohibiting such a continuancy discrepancy above the faithfulness conditions requiring retention of nasality and place, it becomes preferable to lose the nasal entirely rather than violate the syntagmatic constraint. This effect can be seen by comparing the case of a nasal–stop sequence (3.25) with that of a nasal–fricative (3.26).

(3.25)   Zoque: prefixal /n–/ before stops

n+[pama]	ICC [CONT]	ICC [VOICE]	FAITH [NASAL]	FAITH [PLACE]
mpama		*!		*
☞ mbama				*
pama			*!	*

(3.26)   Zoque: prefixal /n–/ before fricatives

n+[šapun]	ICC [CONT]	ICC [VOICE]	FAITH [NASAL]	FAITH [PLACE]
ñšapun	*!	*		
ñžapun	*!			
☞ šapun			*	*

Highly ranked IDENTICAL CLUSTER CONSTRAINT [PLACE] would independently rule out any candidates in which the nasal prefix is not articulated at the same place of articulation as the following stop or fricative (e.g. *npama or *nšapun). Since the effect of this constraint has already been illustrated in (3.23), all the candidates included in (3.25) and (3.26) respect the place constraint. Of interest is the effect of the IDENTICAL CLUSTER CONSTRAINT [CONTINUANCY]. When the nasal is followed by a stop, both the nasal and the stop share a single value for continuancy, namely, both are stops. In contrast, when the nasal is followed by a fricative, it is impossible for the two consonants

---

[4]These three possibilities do not exhaust the range of options available in such a configuration. For additional discussion, see Padgett (1994).

to share continuancy values since the nasal is a stop but the fricative is a continuant. The consequence of this difference is that nasal–stop sequences are acceptable (*mbama*), whereas nasal–fricative sequences are ruled out (*ñšapun, *ñžapun), replaced by a much less faithful form that satisfies the cluster condition on continuancy (*šapun*).

In some cases, a particular sequence of segments may be so undesirable that the best output eliminates one segment entirely.

There is more to the analysis of nasal–fricative clusters in Zoque, however. When a nasal–fricative sequence occurs entirely within a stem, it is retained without modification: *winsaʔu* 'he received', *ʔaŋsis* 'lips', *woʔmsoŋ* 'quail', etc. The nasal neither deletes nor does it assimilate in place or continuancy, nor does the fricative voice. Two additional factors must be considered to account for this fact. First, the stem versus nonstem distinction again plays a role. Faithfulness in the stem domain takes precedence over the IDENTICAL CLUSTER CONSTRAINT [CONTINUANCY]. As a result, nasal deletion can take place when the nasal is not part of the stem, but is blocked if the nasal is stem material. Second, we see that just as nasals in Zoque are invariably stops and invariably voiced, fricatives are invariably voiceless. This type of segment–internal pattern of feature cooccurrence depends on a paradigmatic prohibition against voiced fricatives, the type of constraint discussed in the next section.

4 Deviation from Identity: Paradigmatic Constraints

In the preceding section, we examined cases in which an output deviated from an input in order to obey a constraint governing sequences of sounds. A similar sort of deviation can be observed in cases in which one feature imposes a condition on another feature within the same speech sound.

Recall the Zoque pattern of voicing in nasals. We saw that a sequential constraint requires that a nasal and a following stop share the same value for voicing. Hence an input like /min-pa/, where *n* is voiced and *p* is voiceless, surfaces as [minba], where both *n* and *b* are voiced. We also noted that there is an alternative output that would satisfy the shared voicing requirement, one in which both the nasal and the stop are voiceless: *miŋpa. This particular candidate output can be ruled out by a paradigmatic constraint requiring that nasals be voiced (Itô, Mester, & Padgett 1995).

(3.27)   NAS/VOI: A nasal must be voiced.

Since both *minba* and *miŋpa satisfy the cluster voicing constraint, NAS/VOI intervenes to cause the former to be selected, as in tableau (3.28). The first candidate in (3.28), *minpa, satisfies faithfulness but violates the cluster condition on voicing. The second and third candidates both satisfy the cluster condition, but only the second also satisfies the requirement that a nasal be voiced and is therefore established as optimal.

Note that in (3.28), it has been assumed that the nasal *n* is voiced in the input and therefore that there would be a FAITH[VOICE] violation were it to appear as voiceless in the output, **miŋpa*. Importantly, however, this assumption concerning input values is for expositional explicitness only, playing no role in the establishment of an optimal output. Whether the nasal is voiced, voiceless, or unspecified for voicing in the input, NAS/VOI requires that it be *voiced* in the output.

(3.28) **Zoque: the effect of requiring nasals to be voiced**

/min–pa/	ICC[VOICE]	NAS/VOI	FAITH[VOICE]
minpa	*!		
☞ minba			*
miŋpa		*!	*

NAS/VOI, like other constraints which force deviations from input/output identity, may be ranked either above or below faithfulness.

(3.29) **Possible rankings of NAS/VOI with respect to faithfulness**
   a. NAS/VOI » FAITH
   b. FAITH » NAS/VOI

It has been argued that feature cooccurrence constraints such as NAS/VOI must be motivated phonetically. Such **grounded** cooccurrence constraints are of two types, **sympathetic** and **antagonistic** (Archangeli & Pulleyblank 1994). The sympathetic type positively requires that one feature cooccur with another feature. This is the type that NAS/VOI illustrates: the feature of nasality forces cooccurrence with the feature voicing. Voicing is sympathetic to nasality, enhancing it.

Phonetically motivated paradigmatic conditions require particular patterns of feature cooccurrence within a segment.

Sympathetic   Feature F must appear when feature G appears.

Antagonistic   Feature F must *not* appear when feature G appears.

Of course, in spite of the manner in which voicing enhances nasality, some languages block such enhancement by the high ranking of faithfulness (second case in (3.29)). For example, Burmese contrasts voiced and voiceless nasals — whether a nasal is voiced or voiceless contributes to lexical distinctions in meaning (Ladefoged & Maddieson 1996).

(3.30) **Burmese: minimal pairs showing voiced and voiceless nasals**

VOICED		VOICELESS	
mă	'hard'	m̥ă	'notice'
nă	'pain'	n̥ă	'nose'
nʷă	'cow'	n̥ʷă	'peel'

The only difference between the words *mă* and *m̥ă* is that the initial nasal consonant of *mă* is voiced while the initial nasal of *m̥ă* is voiceless. For such a language, faithfulness to underlying representations must outrank NAS/VOI so as to preserve the voicelessness of the second class of nasals.

Numerous feature relationships are of the sympathetic type, for example: voicelessness with high pitch, voicing with low pitch; tongue body backing with lip rounding; tongue body raising and fronting with tongue root advancement, tongue body lowering and backing with tongue root retraction; stridency with continuancy; and so on. In all such cases, languages vary as to the rigor with which the condition is imposed, the degree of rigor imposed corresponding to where the constraint is ranked in the grammar of a particular language.

In other cases, feature relationships may be antagonistic: one value may be prohibited from cooccurring with another. Some antagonistic relationships are simply the inverses of sympathetic ones. For example, tongue body fronting may not cooccur with lip rounding; tongue body lowering and backing may not cooccur with tongue root advancement, and tongue body raising and fronting may not cooccur with tongue root retraction. Other antagonistic relationships are not related to sympathetic ones. For example, an articulation with the lips may not cooccur in a consonant with an articulation of the tongue tip or blade, nor may a tongue tip or blade articulation cooccur with a tongue body articulation.

To illustrate an antagonistic feature relation, consider again the example of nasality. The most frequent type of nasal segment is a nasal stop — a stop because there is a complete blockage of air in the oral cavity, with nasality resulting from opening of the nasal cavity. The sorts of nasal consonants discussed in Section 3 (*m, n, ŋ*, etc.) are nasal stops. It is also possible, however, to find nasal continuants. The most common are nasalized vowels, as in French *bon* [bɔ̃] 'good', with nasalized liquids, glides and fricatives occurring more rarely. The rarity of such segments can be attributed to an antagonistic constraint NAS/CONT.

(3.31)   NAS/CONT: A nasal must not be continuant.

When this constraint is highly ranked, nasal continuants are ruled out (3.32a); when FAITH dominates NAS/CONT, nasalized continuants become possible (3.32b).

(3.32)   **Possible rankings of NAS/CONT with respect to faithfulness**
   a.   NAS/CONT » FAITH
   b.   FAITH » NAS/CONT

So, a Kuliak language spoken in northeastern Uganda (Carlin 1993), is a language in which NAS/CONT is ranked above FAITH. The only nasal sounds found are the stops [m, n, ñ, ŋ], as in [mad] 'pumpkin', [nɛbɛc] 'two', [ñet] 'men', and [ŋal] 'sharp'. An example of a language that contrasts oral and nasalized continuants is Waffa, a language of New Guinea (Ladefoged & Maddieson 1996).

(3.33)   **Waffa: pairs showing oral and nasalized fricatives**

Oral	Gloss	Nasalized	Gloss
óoβə	'type of yam'	β̃atá	'ground'
βaíni	'close by'	jaáβ̃ə	'nose'

The Waffa pattern is derived by ranking faithfulness above NAS/CONT: it is more important for the output to respect input distinctions concerning continuancy and nasality (FAITH) than to avoid violating the prohibition on combining nasality with fricatives (NAS/CONT).

While feature cooccurrence restrictions, both sympathetic and antagonistic, govern a wide variety of phonological processes, one area in which they play a considerable role is in the determination of segment inventories. We discuss this function in the next section.

5 Constraints on Inventories

In the preceding sections, we have assumed that the words of a language are constructed from various combinations of segments. Typically, when undertaking a description of a language, one of the first goals a linguist sets is to determine the language's sound inventory. Some sample consonant inventories are given in (3.34).[5]

(3.34)   **Sample consonant inventories**

Hawaiian: {p,k,ʔ,h,m,n,l,w}

Japanese: {p,t,k,b,d,g,s,h,z,r,m,n,w,y}

Yoruba: {b,m,f,t,d,s,l,r,š,ǰ,y,k,g,kp,gb,w,h}

English: {p,b,t,d,k,g,f,v,θ,ð,s,z,š,ž,č,ǰ,m,n,ŋ,h,y,r,l,w}

Zoque: {p,t,tʸ,č,k,b,d,dʸ,ǰ,g,c,ʔ,f,s,š,h,m,n,ñ,ŋ,l,ɾ,r,w,y}

Bura: {p,ps,pš,pʷ,pt,pts,pč,t,ts,č,c,k,kʷ,b,bz,bʷ,ʔb,ʔbʷ,ʔbd,d,ʔd,dz,
ǰ,ɟ,g,gʷ,m,mʸ,mʷ,mnpt,n,ñ,ŋ,ʌ,f,fʷ,w,ʔw,w⊥,v,s,š,ç,h,z,ž,y,ʔy,ɣ,
ɬ,ʎ,l,ɮ,ʌ,r}

Obviously such inventories differ considerably in the number of consonants they exhibit, in the particular consonants included in each inventory, and in the complexity of the consonants included. Nevertheless, examination of large numbers of inventories demonstrates that the selection of consonants is not haphazard. Both across languages and within languages, regular patterns emerge. At issue, therefore, is how to express such patterns.

Within derivational frameworks, the precise mechanisms by which underlying inventories are derived are often only briefly dealt with. One approach is to impose a set of

---

[5]For an explanation of how the various symbols included here are pronounced, and for an introductory discussion of their phonetic properties, see Ladefoged (1975). Languages included here are selected from those found in Comrie (1987), supplemented by some of the languages discussed in this work. Sources other than Comrie (1987) are Ladefoged (1968) for Bura, Wonderly (1951) for Zoque.

cooccurrence conditions on the appearance and combination of features in underlying representations (e.g. Kiparsky 1985, Archangeli & Pulleyblank 1994). This proposal can be supplemented by distinguishing between features that are present in all languages and features that are only present in inventories of specific languages (Christdas 1988). Output forms are derived by the application of various rules to the underlying segments so defined; these may eliminate segments (**neutralization**) or add new ones (**allophony**). The overall picture is one in which inventories result from the interaction of input constraints, rules, and output constraints.

In contrast, Optimality Theory hypothesizes that there are no constraints of any kind on input forms.[6] Whether some segment is included in the inventory of a given language depends fully on the nature of the constraint grammar of the language in question. That is, the inventory derives from the way that constraints on output forms interact with freely chosen input feature combinations (Prince & Smolensky 1993; Smolensky 1993).

Consider, for example, the two feature cooccurrence conditions on nasality discussed in the last section, namely NAS/VOI (3.27) and NAS/CONT (3.31). These constraints require that nasals in the output of a language be both voiced (NAS/VOI) and stops (NAS/CONT). This is indeed the case in all of the inventories of (3.34). By ranking these constraints above faithfulness, the output results are assured without any input restrictions. The central point is that the interaction of featural constraints with faithfulness *derives* inventories.

According to Optimality Theory, the content of lexical inputs is unconstrained. Whether some segment occurs on the surface in a particular language is determined strictly by the constraint grammar of the language in question.

If faithfulness to a particular feature outranks any prohibitions governing the appearance of the feature, then the feature contributes to defining a language's inventory.

If prohibitions against some feature outrank relevant faithfulness constraints, then the feature does not play a role in the inventory.

Neutralization and Allophony

Even though a particular feature combination is allowed in some language, this does not necessarily mean that the combination in question will be allowed in all contexts. Consider the case of voicing on obstruents. (Recall that obstruents refer to the class of oral

---

[6]This statement needs to be qualified somewhat. Inputs may only consist of legitimate phonological material — for example, handclaps and winks could not be part of the lexical input — and all such material must obey certain basic and universally respected properties of phonological well–formedness. For example, coronality (use of the tongue tip or blade in an articulation) cannot be represented as a feature of the lower lip in some language, either underlyingly or on the surface. The hypothesis of 'no constraints on input forms' is that none of the *violable* constraints can be imposed as a requirement of inputs. Of course, specific inputs will typically respect at least some constraints, but they are not required to by the theory.

stops and fricatives, such as the voiceless series *p, t, k, s, š* and the voiced series *b, d, g, z, ž*.)

In languages such as English, voiced and voiceless obstruents are **contrastive**. Pairs such as *pan/ban, tent/dent, kill/gill,* and *sip/zip* mean different things and differ from each other phonologically only in the value of voicing present on the initial obstruent. Similar pairs can be distinguished on the basis of medial or final obstruents, for example, *ripping/ribbing, bussing/buzzing, fat/fad, lock/log*. (Note that it is the sounds that are crucial here, not the spelling.) Since the voicing differences in such pairs are lexically idiosyncratic, such voicing distinctions must be encoded in the lexical entries of English words. To ensure that such input distinctions are preserved in the output is the role of the faithfulness conditions.

In many languages, however, all obstruents are voiceless. The voicelessness is redundant. This is the case in Hawaiian, for example. As seen in the inventory of (3.34) and the examples of (3.35), the two obstruents in Hawaiian are voiceless (Pukui & Elbert 1957).

(3.35)   **Hawaiian obstruents: [p, k]**

Word	*Gloss*
kaikea	'sap, sapwood'
kapu	'taboo'
pekapeka	'tattle, tell tales'
popopo	'rot, as of wood or cloth'

The tendency for obstruents to be voiceless derives from the phonetic fact that it is more difficult to maintain vibration of the vocal cords when there is a constriction of the type that produces a fricative or an oral stop. The phonological constraint involved can be referred to as the "obstruent/voice" constraint. Just as "nasal/voice" requires nasals to be voiced, "obstruent/voice" require that obstruents be voiceless.

(3.36)   OBS/VOI: An obstruent must be voiceless.

Whether or not voiced obstruents are possible in the inventory of a language will depend on where OBS/VOI is ranked with respect to faithfulness. For a language like English, in order to ensure that lexical voicing contrasts override the constraint that obstruents be voiceless, it is crucial that FAITH[VOICE] be ranked above OBS/VOI.

(3.37)   **Voiced obstruents are attested in the inventory**
FAITH[VOICE] » OBS/VOI

Given this ranking, it is less important to avoid voicing on obstruents than it is to preserve any instance of voicing that happens to be included in a lexical input representation. Hence both voiced and voiceless obstruents will appear in some outputs, as is indeed the case in languages like English.

In languages like Hawaiian(3.34), however, OBS/VOI is systematically true, defining the inventory. This result is achieved by inverting the relative ranking of faithfulness with the condition on obstruent voicing.

(3.38)   **Voiced obstruents are excluded from the inventory**
         OBS/VOI » FAITH[VOICE]

Note that even in a language like Hawaiian, there are two logical types of input forms with regard to voicing: an input could in principle contain one or more instances of the voicing feature on an obstruent, or an input could contain no instances of voiced obstruents. Given the ranking in (3.38), whether or not an input contains the voicing feature, it must not exhibit it in the output since the prohibition against voicing on obstruents (OBS/VOI) is more important than the requirement to have an output that is identical to the input. The net result is that voiced obstruents would not occur in such a language. Of course, the complete absence of voiced obstruents on the surface would render the postulation of voicing on obstruents in an input form redundant (see Chapter 1, Prince & Smolensky 1993, and Itô, Mester, and Padgett 1995 on **lexicon optimization**). The point is, however, that the absence of voiced obstruents is a result of the grammar fragment in (3.38), not the result of some constraint on the alphabet of phonological inputs.

Neutralization

Consider now a case where a language allows a particular feature combination in general, but disallows it in some specific context. We illustrate this possibility through another example involving voicing on obstruents.

In Russian (Kenstowicz & Kisseberth 1979), voicing is contrastive on obstruents. That is, as in English, the voicing distinction on obstruents leads to differences in lexical meaning. Examples can be seen in (3.39), voiced obstruents to the left, voiceless obstruents to the right. In these examples, syllable boundaries are indicated with a period (e.g. *tru.pu* has two syllables, *tru* and *pu*). The example consonants all come immediately before the dative singular suffix *–u*.

(3.39)   **Russian voicing contrast: examples with the dative singular**

Voiced input	Dative singular	Gloss	Voiceless input	Dative singular	Gloss
/b/	xlé.bu	'bread'	/p/	tru.pu	'corpse'
/d/	sa.du	'garden'	/t/	za.ka.tu	'sunset'
/g/	ro.gu	'horn'	/k/	ra.ku	'crayfish'
/z/	ra.zu	'time'	/s/	le.su	'forest'
/ž/	sto.ro.žu	'guard'	/š/	du.šu	'shower'

Unlike English, Russian does not maintain the voicing contrast in all positions. Specifically, the distinction between voiced and voiceless obstruents is lost at the end of a syllable, where all obstruents appear as voiceless. The distinction is said to be neutralized syllable–finally. This can be seen by comparing the forms of (3.39) with those of (3.40) in the nominative singular. Whereas the dative singular forms are marked by the suffix *–u*, the nominative singular has no affix at all. As a result, all of the root–final consonants appear at the end of a syllable and are therefore voiceless — no matter whether they are voiced or voiceless in the input.

In terms of the required constraint ranking, such behavior at first appears contradictory. Like English, Russian maintains voicing distinctions in certain environments (at the beginning of a syllable), but like Hawaiian, it enforces voicelessness on obstruents in others (at the end of a syllable).

(3.40)   **Russian voicing neutralization: examples with the nominative singular**

Voiced input	Nominative singular	Gloss	Voiceless input	Nominative singular	Gloss
/b/	xlep	'bread'	/p/	trup	'corpse'
/d/	sat	'garden'	/t/	za.kat	'sunset'
/g/	rok	'horn'	/k/	rak	'crayfish'
/z/	ras	'time'	/s/	les	'forest'
/ž/	sto.roš	'guard'	/š/	duš	'shower'

Consider the grammar necessary to achieve such a pattern. If the grammar consisted solely of the ranking of FAITH[VOICE] over OBS/VOI, as in English (3.37), then voiced and voiceless obstruents should be possible in all positions. If the grammar consisted of the opposite ranking, as in Hawaiian (3.38), then voiced obstruents should be completely impossible. What is needed is a third possibility, one that can be derived by introducing an additional constraint that interacts with the two considered so far.

It has been noted in the linguistic literature that codas make poor hosts for a number of features: voicing, place distinctions, articulations involving more than one articulator, and so on. Investigating this property of codas, and formally encoding it, goes far beyond the scope of this chapter. For present purposes, I will informally encode this property as a constraint which requires that codas be minimally specified.

(3.41)   CONTRASTIVECODA: A coda does not bear contrastive features.

This constraint often comes into direct conflict with the requirements of faithfulness. Where a lexical entry happens to have some feature specification such as voicing that ends up in a coda, faithfulness requires that the specification be retained; CONTRASTIVECODA, on the other hand, requires that the feature be lost.

For a language like English, CONTRASTIVECODA must be ranked below FAITH[VOICE] since a *voiced* feature on an input is retained even in a coda.

(3.42)   **English: voiced obstruents are attested even in codas**
         FAITH[VOICE] » CONTRASTIVECODA, OBS/VOI

Note that the relative ranking of CONTRASTIVECODA and OBS/VOI is not crucial in the English case since all outputs must be fully faithful to the corresponding input.

In Russian, on the other hand, CONTRASTIVECODA must outrank FAITH[VOICE], bringing about the neutralization of voicing distinctions in codas.

(3.43)   **Russian: voiced obstruents are impossible in codas**
         CONTRASTIVECODA » FAITH[VOICE] » OBS/VOI

By ranking FAITH[VOICE] above OBS/VOI, as in English, general retention of voicing contrasts is ensured. By ranking CONTRASTIVECODA above FAITH[VOICE], as in Russian, the specific loss of voicing contrasts in codas is achieved. Note that the redundant realization of obstruents as voiceless *emerges* through the interaction of OBS/VOI with the loss of contrastive values resulting from the more highly ranked CONTRASTIVECODA.

The English and Russian patterns are illustrated by the two tableaux in (3.44) and (3.45), respectively. Words with very similar phonological shapes have been used to facilitate the comparison of English and Russian.

(3.44)   **English: general faithfulness to input voicing**

/sad/ 'sod'	FAITH[VOICE]	CONTRASTIVECODA	OBS/VOI
☞ sad		*	*
sat	*!		

In English, retaining the input value of voicing on the final consonant of [sad] is necessary to satisfy the highly ranked FAITH[VOICE], even though this results in violations of CONTRASTIVECODA and OBS/VOI.

In Russian, the high ranking of CONTRASTIVECODA prohibits the appearance of a voiced specification in the coda. The result is that FAITH[VOICE] must be violated in the optimal candidate [sat].

(3.45)   **Russian: neutralization of voicing in codas**

/sad/ 'garden'	CONTRASTIVECODA	FAITH[VOICE]	OBS/VOI
sad	*!		*
☞ sat		*	

In summary, we have seen that inventories allow or disallow particular feature combinations by the ranking of faithfulness conditions relative to cooccurrence conditions. High ranking of faithfulness conditions results in the appearance of a segment showing marked feature combinations; low ranking of faithfulness results in a less marked inventory. In addition, the occurrence of marked segments may be possible in general, but prohibited in some specific context. Such cases are achieved by ranking faithfulness above the general cooccurrence condition, but having a third constraint overrule faithfulness in some restricted environment.

Allophony

We have seen three sorts of cases in this section so far. First, a particular feature is prohibited in all contexts from cooccurring with some other feature, for example, voicing with obstruents; the feature is **noncontrastive**. Second, such a prohibition is systematically overruled by faithfulness to lexical entries; the feature is contrastive. Third, even though the feature is contrastive in general, in some context, the prohibition reasserts itself; there is neutralization. As a final example of constraint interaction, a feature may be noncontrastive, but with a distinction nevertheless arising in a predictable context, a case of allophony.

A case of this type is seen in various dialects of Quechua — I have chosen Quechua because of its similarity to others discussed here. In Imbabura Quechua, a language of northern Ecuador (Cole 1982), there are three voiceless stops: [p, t, k].[7] Except for a class of words borrowed from Spanish, voiced stops are not found contrastively in Quechua.[8] Unlike English and Russian, Quechua does not distinguish words of different meaning by virtue of whether an obstruent is voiced or voiceless. Like Hawaiian, words normally exhibit voiceless stops. For example, the three suffixes –*pi*, –*ta*, and –*ka* in (3.46) illustrate the oral stops of Quechua, and all are voiceless.

(3.46) **Quechua: voiceless stops**[9]

Stops	Words	Gloss
/p/	wasi uku–pi	'inside the house'
	house interior–in	
/t/	marja–ta	'Maria–ACC'
	Maria–ACC	
/k/	marja–ka	'Maria–TOPIC'
	Maria–TOPIC	

Unlike the examples in (3.46), stops in Imbabura Quechua are voiced when appearing after a nasal — exactly as seen for Zoque in Section 3. Hence the same suffixes seen to begin with voiceless stops in (3.46) appear with voiced stops when they are attached to roots ending in nasals, for example, *ñan* 'road'. (Note also that the nasal undergoes place assimilation when followed by /p/; the assimilation is comparable, though not identical, to the processes of assimilation seen in Section 3 and is not discussed here.)

The general pattern of voicelessness for obstruents requires a ranking of OBS/VOI above FAITH[VOICE]. However, ICC[VOICE] must be ranked in turn above OBS/VOI, causing voicing of stops where necessary to satisfy the cluster voicing constraint.

(3.47) **Quechua: postnasal voicing**

Stops	Words	Gloss
/p/	ñam–bi	'in the road'
	road–in	
/t/	ñan–da	'road–ACC'
	road–ACC	
/k/	ñan–ga	'road–TOPIC'
	road–TOPIC	

To see this, consider a form such as *marja–ta* 'Maria–ACC'. For the general case, since voiced stops do not independently occur in Quechua, it must be the case that whatever the input, the optimal output of a stop is voiceless. This results from ranking of OBS/VOI

---

[7] There are also affricates in Quechua ([ts, č]) which are not discussed here.
[8] Given the high rate of bilingualism between Quechua and Spanish, and the number of borrowings from Spanish that include voiced stops, it may now be more appropriate to analyze Quechua synchronically as having a contrast between voiced and voiceless stops. If this is the case, then the pattern of voicing described here would correspond to a variety of Quechua prior to such contact.
[9] The effect of vowel laxing is not indicated in the Quechua transcriptions.

above FAITH[VOICE]. For example, if the input stop is voiceless, then the optimal output is *marja–ta*, a form that both respects OBS/VOI and is fully faithful.

(3.48)   **Imbabura Quechua: voiceless stop in input**

/marja–ta/	OBS/VOI	FAITH[VOICE]
☞ marja–ta		
marja–da	*!	*

If the input were to posit a voiced stop, i.e. /marja–da/, then *marja–ta* would still be derived.

(3.49)   **Imbabura Quechua: putative voiced stop in input**

/marja–da/	OBS/VOI	FAITH[VOICE]
☞ marja–ta		*
marja–da	*!	

However, this requirement of the grammar to produce voiceless stops is overridden by the effect of the high–ranking constraint that requires consonant clusters to share a single specification for voicing. The tableau in (3.50) shows the effect of adding ICC[VOICE], the cluster constraint requiring that all consonants in a cluster share a single voicing specification (Section 3), to the set of constraints governing voicing. (The constraints governing place of articulation have not been included in (3.50) since the issue of place agreement is orthogonal to that of voicing.)

(3.50)   **Imbabura Quechua: overriding requirement that obstruents be voiceless**

/ñan–ta/	ICC[VOICE]	OBS/VOI	FAITH[VOICE]
ñan–ta	*!		
☞ ñan–da		*	*

Regardless of the fact that OBS/VOI rules out voiced stops in general, ICC[VOICE] forces their occurrence after a nasal. (Also, the paradigmatic constraint NAS/VOI would have to be highly ranked to ensure that the nasal remains voiced.)

In conclusion, the distinction between voiced and voiceless stops is *allophonic* in Imbabura Quechua, meaning that the distribution of the two values is predictable and complementary: a voiced stop occurs after a nasal, a voiceless stop occurs elsewhere. Within Optimality Theory, this type of pattern can be derived by the interaction of three constraints. A feature cooccurrence constraint dominates faithfulness, removing the sound from the language's general inventory; some more specific constraint then forces the feature's appearance in some specific context.

The distribution of a feature

Feature *F* occurs in the inventory: FAITH[F] » *F

Neutralization: *F* occurs in an inventory, but a context–specific condition overrides general considerations of faithfulness:

SPECIAL CONDITION ON F » FAITH[F] » *F

Feature *F* does not occur in the inventory: *F » FAITH[F]

Allophony: *F* does not occur in an inventory, but a context–specific condition overrides the general prohibition:

SPECIAL CONDITION ON F » *F » FAITH[F]

---

6 Deviation from Identity: Constraints on Association

Up to this point, lexical inputs have been presented as though they always consist of complete speech sounds. For example, the meaning of *cat* is paired with the string of speech sounds /kæt/, where the sound *k* consists of the various features that make up *k*, the sound *æ* consists of the various features that make up *æ*, and so on. In many languages, however, there are morphemes whose content may be incompletely specified segments, even single features (for recent accounts and references, see Archangeli & Pulleyblank 1994, and Akinlabi 1995, to appear). For example, in Terena, a Southern Maipuran language of Brazil (Akinlabi to appear), the word for 'brother' is *ayo*. To form the meaning 'my brother', a single feature is added, the feature of nasality: *ãỹõ*. No full segment is added to mean 'my' — nasality represents that meaning.

Such cases may involve numerous different features. For example, Chaha, a South Semitic language of Ethiopia, indicates feminine imperatives by the addition of a palatalization feature, and third person masculine objects by the addition of labialization (McCarthy 1983). Tiv, a Niger–Congo language of Nigeria represents the general past tense by the addition of a low tone, and the recent past tense by the addition of a high tone (Arnott 1964, Pulleyblank 1986). Igbo, also a Niger–Congo language of Nigeria, represents the possessive marker by a high tone, with the precise realization of the tone varying from dialect to dialect (Welmers 1970, Hyman 1975).

To illustrate, consider the Igbo case. In Central Igbo, the words in (3.51) are pronounced as indicated when they occur in isolation; *á*, *é*, etc. indicate vowels with a high tone, *à*, *è*, etc. indicate vowels with a low tone.

In Igbo, meanings are distinguished by virtue of the pitch of a vowel. The use of pitch to contrast different meanings is referred to as **tone** (as mentioned briefly in Section 3). For example, the meanings of the words *ákwá*, *àkwà*, *àkwá*, and *ákwà* are solely differentiated by their tones: the word *ákwá* has two high tones, *àkwà* has two low tones, *àkwá* has a pattern of low tone followed by high tone, while *ákwà* has high tone followed by low.

(3.51)   Igbo: words in isolation

Word	Gloss
àgbà	'jaw'
èŋwè	'monkey'
ákwá	'crying'
àkwà	'bed'
àkwá	'egg'
ákwà	'cloth'

When the first two words of (3.51) are combined to form a possessive phrase, 'jaw of monkey', they are put together in order àgbà + èŋwè, and the low tone on the final vowel of àgbà is replaced by a high tone (àgbá).

(3.52)   Central Igbo: possessive phrase
        àgbá èŋwè   'jaw of monkey'

Recall from Section 2 that an input representation consists of speech sounds paired with a meaning in the mental lexicon. As such, the input to a phrase such as (3.52) would be as in (3.53), where L represents low tone and H high tone. Note that no phonological segment is associated with the high tone of the possessive morpheme.

(3.53)   Central Igbo: input representation

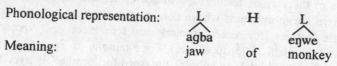

Phonological representation: L   H   L

Meaning: agba   eŋwe
     jaw   of   monkey

If the output for such an input were to completely satisfy faithfulness, then it would be identical to the input. As such, no vowel would actually bear the high tone of the possessive morpheme, and the surface tonal pattern would consist of four vowels, all with low tones. Note that such a faithful output would result in the elimination over time of the high tone possessive marker. Since a child learning the fully faithful variety of Igbo would not encounter any reflex of the high tone of the possessive marker, such a generation of speakers would not learn it.

The actual output deviates from such a faithful representation by virtue of the appearance of high tone on the final vowel of the first word in the phrase:

(3.54)   Central Igbo: output representation

L   H   L
| | /\
agba   eŋwe

By virtue of its association to a vowel, the high tone is *linked* to a vowel, and therefore pronounced. The LINK[TONE] constraint is satisfied:[10]

---

[10]Such 'linking' is also referred to as 'parsing' in the optimality literature. The term 'link' is used here to avoid confusion with a variety of other uses of the term 'parse'. Note in particular that

(3.55)   LINK[TONE] : A tone must be associated with a vowel.

For Igbo, associating the high tone of the possessive is more important than being entirely faithful, that is, LINK [TONE] outranks FAITH.

(3.56)   Igbo: LINK[TONE] » FAITH

The effect of this ranking is illustrated in the tableau in (3.57). In the first candidate, *àgbà èŋwè, the high tone of the possessive marker is not linked; in the second and winning candidate, àgbá èŋwè, the high tone is linked, even at the expense of a faithfulness violation.

(3.57)   Central Igbo: the necessity of associating a high tone

/àgbà + H + èŋwè/	LINK [TONE]	FAITH
[àgbà èŋwè]	*!	
☞ [àgbá èŋwè]		*

The crucial point of an example such as (3.57) is not *where* the high tone of the possessive associates, but simply *whether* it associates. This can be seen clearly by comparing the possessive phrase seen above in Central Igbo with the corresponding phrase in Aboh Igbo. In Aboh, the word for 'jaw' is *èbgà* and the word for 'monkey' is *èŋwè*. When joined in a possessive phrase, the result is *ègbà éŋwè* 'jaw of monkey', as illustrated in (3.58).

(3.58)   Aboh Igbo: possessive phrase

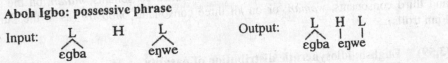

Note that where Central Igbo associates the high tone of the possessive marker to the final vowel of the first noun, Aboh Igbo associates the same marker to the initial vowel of the second noun.

As discussed in Section 5, Optimality Theory hypothesizes that input representations are unconstrained, that the properties of outputs result from the effect of constraints on freely idiosyncratic lexical forms. One type of lexical idiosyncrasy is the inclusion in an input of unassociated featural material — features of nasality, palatalization, tone, and so on. Faithfulness imposes the requirement that a feature appear in the output exactly as in the input. For an unassociated feature, this would mean that if a feature happened to be unassociated in the input, then it would remain unassociated in the output. Constraints of the LINK family intervene to prevent a feature from remaining unlinked. Whatever the status of a feature in the input, LINK requires that the feature be associated to a consonant or a vowel, in the output.

---

though 'PARSE' (defined somewhat differently) is a member of the faithfulness family in work such as Prince & Smolensky (1993), McCarthy & Prince (1993a), LINK is not a faithfulness constraint.

> The LINK family of constraints requires that a feature be associated to a consonant or a vowel, whether or not such an association is part of the lexical input.
>
> Some morphemes introduce unassociated features: LINK causes such features to associate.

7 Deviation from Identity: Constraints on Distribution

While the LINK constraint forces a feature to be linked, there are various ways in which the linking could be achieved. For example, the feature could be linked to the beginning of a domain (e.g. the vowel feature [+low] in Tiv), to the end of a domain (e.g. labialization in Chaha), or the feature could distribute itself throughout a domain (e.g. vowel harmony in Yoruba). All such distributional constraints are in opposition to faithfulness, which would require that whatever the idiosyncratic distribution of some input, that distribution would also be found in the output.

For a case where the distribution of a feature is completely haphazard, consider the distribution of nasality in English. Given a word with three consonants, it is possible for none of the consonants to be nasal, *petal*; nasality may appear on the first consonant only, *metal*; on the second consonant only, *panel*; on the third consonant only, *button* (note that the letters "tt" refer to a single consonant in this and similar cases); on the first and second consonants, *minute*; on the first and third consonants, *mitten*; on the second and third consonants, *woman*; or on all three consonants, *moomin* (as in the family of Finn trolls).

(3.59)   **English: idiosyncratic distribution of nasality**

N	N	N	N N	N N	N N	N N N
pɛtəl	mɛtəl	pænəl	bʌtən	mɪnət	mɪtən	wumən	mumən
petal	metal	panel	button	minute	mitten	woman	moomin

There is nothing predictable about the distribution of nasality in English consonants — nasality is simply an idiosyncratic property of individual consonants, whatever their position in a word. Phrased in terms of constraints, it is crucial that whatever universal conditions constrain the distribution of nasality within some domain in other languages, those constraints are outranked by faithfulness in English.

In contrast to the English pattern of high-ranking input-output identity, other languages impose stringent distributional constraints on particular features. As mentioned, some languages require that certain features appear at either the left edge or the right edge of a domain such as the root or stem. Such a distribution can be achieved by the postulation of **alignment** constraints. Following the proposals for alignment in McCarthy & Prince (1993b), such constraints require that the span through which a feature extends be aligned either with the right edge of a domain, with the left edge of a domain, or with both edges of a domain.

As illustration, consider the pattern of tone distribution in Margi, a Chadic language of Nigeria (Hoffmann 1963, Pulleyblank 1986, 1993 and references therein). Margi, like Igbo contrasts low and high tones. For example, *fwà* 'pool' is pronounced with a low tone while *fwá* 'to put (one) into' is pronounced with a high tone. Similarly, both vowels of *kùbà* 'a white guinea–corn' are low toned, while both vowels of the minimally different *kúbá* 'to pass by' are high toned. Also as in Igbo, the tones on consecutive vowels in Margi may differ. Hence *kábì* 'a small tree' has a high–low pattern, while *kàbí* 'before' has the pattern low–high.

Of particular interest are the patterns observed with verb roots and suffixes. Verb roots themselves may be low or high. For example, *ptsà* 'to roast' is pronounced with a low tone, while *tá* 'to cook' is pronounced with a high tone. A suffix may also have an inherent tone. For example, the suffix *–bá* , which indicates thoroughness or completeness (amongst other things), bears a high tone. The effect of attaching *–bá* to roots bearing low and high tones is unsurprising, the result being low–high and high–high patterns, as seen in (3.60).

(3.60) **Margi: high tone suffixes with low and high roots**

	Root	Gloss	Root + suffix	Gloss
Low tone roots	ptsà	'roast'	ptsàbá	'roast thoroughly'
	pcì	'wash'	pcìbá	'wash out'
	tsàvà	'pierce'	tsàvàbá	'pierce through'
High tone roots	tá	'cook'	tábá	'cook all'
	pá	'build'	pábá	'repair'
	ŋgúlí	'roar'	ŋgúlíbá	'surpass in roaring'

Unlike high tone suffixes such as *–bá* which bear a consistent tone, there are other suffixes whose tones vary depending on the root. Consider, for example, the tonal patterns of words with the suffix *–na*, which typically means that the action of the verb is "done in the direction 'away' or results in some kind of separation, removal, or even destruction" (Hoffmann 1963: p.132).

(3.61) **Margi: changing tone suffixes with low and high roots**

	Root	Gloss	Root + suffix	Gloss
Low tone roots	pcì	'wash'	pcìnà	'wash'
	ghàl	'grow old'	ghàlnà	'wear out'
	tlà	'cut'	tlànà	'cut off'
High tone roots	tá	'cook'	táná	'cook and put aside'
	sá	'get lost'	sáná	'lose'
	pə́r	'bathe'	pə́rná	'take a bath'

An examination of (3.61) shows that the suffix is systematically low after a low root and high after a high root. This distribution can be accounted for by analyzing the suffix *–na* as having no intrinsic tone, requiring that it acquire its tone from the vowel of the abutting root.

(3.62)   **Margi: changing tone suffixes**

    a.   L   b.   H

        pci - na   ta - na

This 'spreading' of the root tone can be achieved by postulating constraints on the alignment of tones. For reasons to be discussed shortly, we focus on the alignment of high tones.

(3.63)   **Distributional constraints**

    ALIGNLEFT[H] : The *left* edge of a high tone span is aligned with the *left* edge of the verb.

    ALIGNRIGHT[H] : The *right* edge of a high tone span is aligned with the *right* edge of the verb.

> The ALIGN family of constraints requires that the domain of a feature extend to the edge of a constituent, for example, the edge of the root or the word.
>
> Cases of 'harmony' or 'assimilation' result when morphemes introduce a feature that is subject to left– or right–edge constraints on its alignment.

For the data in (3.61), the crucial constraint is ALIGNRIGHT[H], which requires that the span through which a high tone extends be aligned with the right edge of the word. A fully faithful output would retain the root tone strictly on the root (the first candidate in (3.64)); an output respecting alignment must violate faithfulness by assigning the root tone to both the root and suffix (the second candidate in (3.64)).

(3.64)   **Margi: spreading of high tone from root to suffix**

H /ta - na/	ALIGNRIGHT [H]	FAITH
H ta - na	*!	
☞ H ta - na		*

Just as suffixes can be toneless in the input, so can roots (it would indeed be problematic if this were not the case, since optimality theory does not posit constraints on inputs). When a toneless root occurs by itself, it is pronounced with a low tone (3.65a); when it occurs with a toneless suffix, both root and suffix are pronounced low (3.65b); when it occurs with a high tone suffix, however, both root and suffix are high (3.65c).

Consider the cases where no tones are present in the input, either on a root by itself or on a root plus any suffixes. Clearly the consistently low tone outputs appropriate for such cases cannot be derived by the alignment of some pre–existing tone since there isn't any.

The options are therefore to leave the representation tonally unspecified or to fill in a low tone.

(3.65)   Margi: toneless roots

		Input	Output	Gloss
a.	*Toneless roots*	fa	fà	'take'
		ɓəl	ɓə̀l	'break'
		tətəl	tə̀tə̀l	'spread'
		ŋal	ŋàl	'bite'
		shu	shù	'stir'
b.	*Toneless roots*	fa + na	fànà	'take away'
	+ toneless suffixes	ɓəl + na	ɓə̀lnà	'break (off)'
		tətəl + na	tə̀tə̀lnà	'spread (something)'
c.	*Toneless roots*	fa + bá	fábá	'take out'
	+ high suffixes	ŋal + bá	ŋálbá	'bite a hole'
		shu + bá	shúbá	'stir well'

The latter option is selected in Margi. First, the low pitch appearing in such "toneless" contexts is apparently indistinguishable from the pitch of a low tone that was present in the input. This is accounted for if the output of the phonology assigns identical representations to the two cases, namely, low tone. Second, there does not appear to be any evidence for gradience or interpolation, the sorts of phonetic markers that would suggest phonetic, rather than phonological, assignment of values. I assume, therefore, that the output appropriate for a form such as *fànà* 'take away' is as in (3.66).

(3.66)   Margi: output where there are no tones in the input

Two issues need to be addressed in this regard. First, what constraint forces the presence of tone in such cases? Second, why low tone, as opposed to some other value?

On the first issue, I assume that there is a constraint requiring that all prosodic units be sufficiently specified to be interpretable. A careful consideration of just what these interpretability requirements are would go well beyond the scope of this chapter, but for present purposes, it suffices to require that all vowels be assigned a tone. This is at the very least phonetically true because all vowels are realized with some pitch, whether high, mid, low, falling, or whatever.

(3.67)   INTERPRETABILITY[TONE]: All vowels must bear a tone.

INTERPRETABILITY and LINK together constitute two thirds of the UNIVERSAL ASSOCIATION CONVENTION (UAC) proposed by Goldsmith (1976). Therefore, they will be referred to jointly as the UAC[11] in cases where it is not necessary to distinguish between the two constraints.

INTERPRETABILITY, if ranked above FAITH, will ensure that the optimal output for a toneless input will be specified for tone.

(3.68)   **Margi: toneless inputs**

/fa – na/	INTERP	FAITH
fa - na	*!*	
☞ L ⌒ fa - na		*

The second issue must now be addressed, namely, why does a low tone in the output satisfy INTERPRETABILITY instead of a high tone, or some other tone. Work on a variety of tone systems suggests that high tones are more likely to be retained than low tones in contexts of deletion, that high tones are more likely to spread than low tones, that low tones are more likely to be malleable in their behavior, and so on (for recent discussion, see Jiang–King 1996). This suggests a harmonic ranking (see Section 3) where the faith-fulness constraints governing high tones are ranked above those governing low tones. (We don't address the issue of low tones vs. mid tones as defaults here, but see Jiang–King 1996.)

(3.69)   Harmonic ranking: FAITH[H] » FAITH[L]

Given the ranking of (3.69), default presence of a low tone is preferred to that of a high tone. This can be seen by expanding the tableau of (3.68) to incorporate the distinction between FAITH[H] and FAITH[L].

(3.70)   **Margi: choosing between default H and default L**

/fa – na/	INTERP	FAITH[H]	FAITH[L]
fa - na	*!*		
☞ L ⌒ fa - na			*
H ⌒ fa - na		*!	

---

[11] The third part of Goldsmith's (1976) UNIVERSAL ASSOCIATION CONVENTION is the constraint that autosegmental associations may not cross. For recent discussion of crossing, see Archangeli & Pulleyblank (1994). I assume here that "no crossing" is an intrinsic part of GEN, such that no candidates violating this condition are ever produced by any grammar. Nothing depends on this assumption in the discussion here.

The appearance of either a low tone or a high tone violates faithfulness when there is no such tone in the input. Since FAITH[H] is ranked above FAITH[L], it is nevertheless worse to insert a high tone than to insert a low tone.

> When INTERPRETABILITY outranks faithfulness, default specifications result; when ranked in the opposite way, phonological outputs may be incompletely specified.
> Default specifications are determined by harmonic feature rankings.

Given the high ranking of INTERPRETABILITY, all output vowels must have a tone. Given the high ranking of the constraints governing high tones, any lexical high tone must be retained in the representation (FAITH[H]) and aligned with the right edge of the word (ALIGNRIGHT[H]).

Note that the analysis presented so far bears a strong resemblance to the sort of analysis possible under a derivational account. Under the derivational account, lexically specified tones associate (cf. LINK) and spread (cf. ALIGNRIGHT[H]), with leftover vowels assigned low tones by default (cf. INTERPRETABILITY interacting with FAITH[H] and FAITH[L]). This resemblance turns out to be more illusory than real, however. There is a fundamental distinction between the two theories, a distinction which turns out to have significant consequences for the analysis of Margi. The distinction lies in the level of representation governed by constraints. To see this, we examine certain gaps in the range of tonal possibilities attested in Margi, and then show how these gaps can be accounted for in both derivational and optimality approaches.

8 The Interaction of FAITH and ALIGN

In this section, we examine four problems of tonal distribution that an account of Margi tone must address. We show that derivational accounts of these patterns require ad hoc stipulations, while an optimality account is possible with the constraints motivated in Section 7 plus a generic constraint prohibiting contour tones.

Problem 1: No Low Tone Suffixes

In the preceding section, three tonal possibilities were expressed for Margi morphemes: namely high, low, and toneless. By combining these three tonal possibilities on roots with the same three tonal possibilities on suffixes, nine possible tonal combinations are predicted, as shown in (3.71). Of the nine predicted patterns, six have been illustrated so far (sample examples are included in (3.71) as a reminder). The remaining three are problematic, however, since they do not appear to exist. That is, there do not appear to be any suffixes that are specified in the input as bearing a low tone. This is a potential problem for any theory not postulating constraints on input tonal representations, and therefore a potential problem for Optimality Theory.

(3.71) **Margi: possible tonal combinations on roots and suffixes**

	Toneless	Low	High
Toneless	L L (fànà)	L L (*none*)	H H (fábá)
Low	L L (pcìnà)	L L (*none*)	L H (pcìbá)
High	H H (táná)	H L (*none*)	H H (tábá)

Problem 2: No Falling Contours

An interesting property of many tone languages is that high and low tones may be combined on a single vowel. If a vowel starts on a low tone and ends on a high tone, a rising contour results; if a vowels starts high and ends low, the result is a falling contour.

(3.72) **Margi: contour tones**
   a. Rising contour   L H   **b.** Falling contour   H L
                               ＼∕   ＼∕
                                a   a

Margi exhibits a pervasive asymmetry with respect to contours: rising contours are possible, but falling ones are not. Hence roots such as those in (3.73) are attested, but comparable roots with falling contours are nonexistent.

(3.73) **Margi: rising contour roots**

Root	Gloss
fĭ	'swell'
hǎ	'grow up'
věl	'fly; jump'
bdlǎ	'forge'
njĭ	'pass (urine)'

Problem 3: No High–Low Patterns in Roots

In addition to the impossibility of falling contours, it is impossible to have a high–low sequence within a Margi verb root. On disyllabic roots, a low–high pattern is possible, in roots such as those in (3.74), but corresponding high–low roots are completely unattested.

(3.74) **Margi: low–high roots**

Root	Gloss
ghàɗá	'get tired'
ŋgùshí	'laugh'
zàɗá	'take off'

Problem 4: Contour Simplification

When a root with a rising contour appears with a suffix, the contour tone is lost. When the suffix is toneless, the low–high sequence that composes the contour is distributed over the root–suffix sequence.

(3.75) **Margi: rising contour roots with toneless suffixes**

Root	Gloss	Root + suffix	Gloss
bdlǎ	'forge'	bdlə̀ná	'forge'
njǐ	'pass (urine)'	njìná	'pass (urine)'
vǎl	'fly; jump'	vəlná	'fly'

When the suffix is high, the rising tone is simplified to a level low tone. (Recall that there are no low tone suffixes, Problem 1.)

(3.76) **Margi: rising contour roots with high tone suffixes**

Root	Gloss	Root + suffix	Gloss
fǐ	'swell'	fìbá	'make swell'
hǎ	'grow up'	hə̀bá	'rear'
vǎl	'fly; jump'	və̀lbá	'jump over, across'

Output vs. Input Constraints, and Violability

Four properties have just been outlined, (i) that there are no low tone suffixes, (ii) that there are no falling contours, (iii) that there are no high–low sequences within morphemes, (iv) that there are no contour tones non–finally. In developing an account of these patterns, it is instructive to compare an optimality theoretic approach to the absence of falling contours and high–low patterns with a derivational account (Pulleyblank 1986).

The basic approach within derivational theory has been to posit **tonal melodies** for such languages, where different languages select different melody sets. For Margi, it would be stipulated that the inventory of tonal melodies is {L; H; LH}, with the additional possibility of leaving a morpheme toneless. With respect to why there are no falling contours, and why there are no high–low sequences, there would be a unified response: there is no {HL} melody. Note, however, that while the melodic account unifies an account of two gaps, it achieves its goal by simple stipulation. A crucial aspect of the melodic account is that it governs input forms. An input with a {LH} melody is well–formed; an input with a {HL} melody is ill–formed. Because of this *input orientation*, a separate stipulation is required for the absence of low tone suffixes. That is, the absence of derived HL sequences, as would occur were a high tone root to be followed by a low tone suffix, cannot be related to the absence of an input melody {HL}. An additional, and formally unrelated, stipulation is therefore required, that low tone suffixes are disallowed. Finally, there would be no reason in principle for rising contours to not precede either a low tone or a high tone, deriving forms like *vəlná or *vəlnà. Either some specific ordering of rules deriving rising contours, or some rule simplifying rising contours in the appropriate context is required.

An optimality account is significantly different for two intrinsic reasons. First, it accounts for the absence of falling contours, as well as derived and underived high–low sequences by exactly the set of constraints already seen. The reason that it is able to do this is its *output orientation*, rather than input orientation. Second, it is able to account for the restricted distribution of rising contours by an extremely general constraint whose effect is to rule out contours in general. Because of the *violability* of constraints within Optimality Theory, such a general constraint is adequate for Margi, ranked appropriately with respect to the set of constraints seen so far.

Since the proposed optimality analysis requires the constraint set and the constraint ranking already motivated for the simple patterns, with the already noted addition of a constraint prohibiting contours, the only thing that this section needs to do is work through various word–types, demonstrating that the analysis already proposed works.

In Section 7, it was shown that the UAC (LINK and INTERPRETABILITY) is highly ranked, requiring that any tones present in a representation be associated (LINK) and that any vowels present be assigned a tone (INTERPRETABILITY). It was also seen that ALIGNRIGHT[H] must dominate faithfulness to derive "spreading", and that within faithfulness, FAITH[H] must be ranked above FAITH[L] to get the effects of "default low insertion". Overall, therefore, the ranking for the tonal grammar of Margi as seen so far is that in (3.77).

(3.77)   UAC » ALIGNRIGHT[H] » FAITH[H] » FAITH[L] » ALIGNLEFT[H]

Given this ranking consider the output result of including an input with a single vowel and a low–high tonal sequence.

The UAC excludes any candidate involving either unlinked tones or toneless vowels, hence the first candidate in (3.78) is ungrammatical. All of the candidates satisfy ALIGNRIGHT[H] since their sole high tone is at the extreme right periphery of the word domain, hence the right–edge alignment constraint excludes none of the candidates in (3.78). The third and fourth candidates in (3.78) are excluded because each is unfaithful to its tonal input, the third by deleting a high tone, the fourth by deleting a low tone. The optimal candidate is therefore *vəl*, the form that correctly includes a rising contour.

(3.78)   **Margi: rising contours**

L H / vəl /	UAC	ALIGNRIGHT [H]	FAITH [H]	FAITH [L]	ALIGNLEFT [H]
L H vəl	*!**				*
☞ L H ⌄ vəl					*
L | vəl			*!		
H | vəl				*!	

Before leaving this example, let us use it to motivate the ranking of the constraint mentioned above that prohibits contour tones.

(3.79) NOCONTOUR: No more than one tone may be linked to a single vowel.

Clearly, since contours are allowed in Margi, the NOCONTOUR constraint must be ranked below the UAC, since it does not override LINK, and it must be ranked below FAITH[H] and FAITH[L], since it does not cause part of a contour to disappear (the second and third candidates in (3.80)). Its ranking with respect to ALIGNLEFT[H] cannot be determined. (Note that "ALIGNR[H]" and "ALIGNL[H]" are used to abbreviate "ALIGNRIGHT[H]" and "ALIGNLEFT[H]" respectively.)

(3.80) **Margi: the ranking of NOCONTOUR**

L H /vəl/	UAC	ALIGNR [H]	FAITH [H]	FAITH [L]	NO CONTOUR	ALIGNL [H]
☞ L H ⌣ vəl					*	*
H | vəl				*!		
L | vəl			*!			

Having established that there is no problem for the control case, the case of a rising contour, let us now consider the effect of the proposed constraint ranking on putative falling contours. Imagine an input with a single vowel and a high–low tonal sequence.

(3.81) **Margi: impossibility of falling contours**

H L /CaC/	UAC	ALIGNR [H]	FAITH [H]	FAITH [L]	NO CONTOUR	ALIGNL [H]
H L CaC	*!**					
H L ⌣ CaC		*!			*	
L | CaC		*!				
☞ H | CaC					*	

The UAC excludes any candidate with unattached tones and unspecified vowels (the first candidate in (3.81)). And crucially different from the case of a rising contour, ALIGNRIGHT[H] excludes any candidate where a high tone is not aligned with the right edge of a word. Hence the second candidate in (3.81), the falling contour candidate, is excluded. Given the impossibility of a falling contour, the general preference for faithfulness towards high tones over low tones would select the fourth candidate in (3.81) over the third candidate. The net result is that given the grammar already motivated for simple alignment cases involving high tones, the exclusion of falling contours is achieved as a by–product.

Exactly the same constraint ranking eliminates the possibility of having a high–low sequence within a morpheme. Compare the treatments of inputs with two vowels and a low–high tonal sequence, (3.82), with that of a comparable input involving a high–low sequence, (3.83). Note that in this and subsequent tableaux for Margi, all candidates considered respect the UAC (both LINK and INTERPRETABILITY); since they would invariably be ruled out, given the high ranking of the UAC, imaginable candidates violating it are not included in the tableaux.

(3.82)   **Margi: low–high sequences are retained**

L H /zədə/	ALIGNR [H]	FAITH [H]	FAITH [L]	NO CONTOUR	ALIGNL [H]
☞ L H ‖ zədə					*
L ∧ zədə		*!			
H ∧ zədə			*!		

In all three candidates, ALIGNRIGHT[H] is satisfied, either because a high tone is aligned with the right edge of the word (the first and third candidates in (3.82)) or vacuously because there is no high tone present (the second candidate in (3.82)). In the two candidates where a tone has been deleted (the second and third candidates), there is a fatal violation of faithfulness. The optimal candidate is therefore the faithful one, (the first candidate in (3.82)).

In contrast, consider the effect of positing an input sequence of high followed by low. In such a case, it is preferable to delete the low tone (the third candidate in (3.83)) — at the cost of a faithfulness violation — rather than misalign the initial high (the first candidate in (3.83)). The optimal candidate, therefore, would be a level high tone sequence — even if the input were to contain a low tone.

It can immediately be noted that exactly the same result obtains if the high–low sequence spans a morpheme boundary. That is, if the ...CaCa... sequence were to consist of two morphemes, ...Ca+Ca..., exactly the same pattern of violations seen in (3.83) would hold: the optimal output would eliminate the low tone. (No tableau is necessary; the reader need simply replace ...CaCa... by ...Ca+Ca... in each candidate.) Hence even if a suffix were assumed to have a low tone in its lexical entry, the result of such a representation would be identical to assuming that the suffix had been tonally unspecified (cf. (3.64)).

To sum up the analysis to this point, it has been demonstrated that the ranking motivated by the simple cases involving single tones also derives those cases involving pairs of tones. A single ranking derives the possibility of rising contours, low–high sequences within morphemes, and low–high sequences across morphemes, as well as the impossibility of falling contours, high–low sequences within morphemes and high–low sequences across morphemes. Optimality Theory is able to unify these potentially disparate patterns by virtue of its output orientation.

(3.83) **Margi: impossibility of a high–low sequence**

H L /CaCa/	ALIGNR [H]	FAITH [H]	FAITH [L]	NO CONTOUR	ALIGNL [H]
H L CaCa	*!				
L CaCa		*!			
☞ H CaCa			*		

As the final point, consider the case of contour simplification. Imagine an input with a rising contour, for example, the root *vəl*, and suppose (to stack the deck most massively against the optimality analysis) that the input representation has a prelinked contour. Suppose, moreover, that toneless and high–toned suffixes are then attached to such a morpheme. The results of such affixation are shown in (3.84) and (3.85). For expository simplicity only, the UAC is not included, and all candidates presented respect the requirements of the UAC.

(3.84) **Margi: rising contour root with toneless suffix**

L H / vəl - na/	ALIGNR [H]	FAITH [H]	FAITH [L]	NO CONTOUR	ALIGNL [H]
L H vəl - na				*!	*
☞ L H vəl - na					*
L vəl - na		*!			
H vəl - na			*!		

For the case of a toneless suffix, FAITH[H] and FAITH[L] rule out candidates where input tones are deleted (the third and fourth candidates in 3.84)). Of the two remaining candidates, NOCONTOUR rules out the form where there is an initial rising contour, hence *vəlná* (the second candidate in (3.84)) is selected as optimal.

This example illustrates an important property of Optimality Theory. Constraints are violable; so even though they may be ranked lowly, they are not eliminated. Allowing or disallowing contour tones is not the same as turning a parameter on or off. If there was a contour parameter whose setting was "on" for Margi in general (so as to allow rising

DOUGLAS PULLEYBLANK

tones), one would be unable to explain the absence of contours in suffixed forms by turning the parameter "off". Within Optimality Theory, if the constraint prohibiting contour tones is violated, this means that it must be relatively lowly ranked, but it is not "off". Given the appropriate circumstances, its effect will be felt — in this case, the prohibition of a potential rising contour. The same prohibition can be seen when a rising contour root is followed by a high tone suffix.

(3.85)   **Margi: rising tone root with high tone suffix**

L H H (vǝl-ba) / vǝl - ba/	ALIGNR [H]	FAITH [H]	FAITH [L]	NO CONTOUR	ALIGNL [H]
L H H — vǝl - ba	*!			*	***
L H — vǝl - ba		*		*!	*
☞ L H — vǝl - ba		*			*
L — vǝl - ba		**!			
H — vǝl - ba		*	*!		

In this case, the completely faithful candidate (the first candidate in (3.85)) is ruled out because of its violation of the high ranking ALIGN RIGHT[H]: even though the suffixal high tone is successfully aligned with the right edge of the word, the root high is misaligned. Retention of both high tones results in an alignment violation, so at least one high tone must be deleted. Notice that no more than one high tone can be deleted since deletion of two highs would cause a fatal FAITH[H] violation (the fourth candidate in (3.85)). No constraint forces deletion of the root low tone, hence the fifth candidate in (3.85) is also nonoptimal. So the ultimate choice between candidates comes down again to the question of whether NOCONTOUR is respected (the third candidate in (3.85)) or violated (the second candidate in (3.85)).

In conclusion, by focusing on outputs, and by doing so with violable constraints, the analysis of Margi presented here straightforwardly resolves the sorts of problems encountered in multi–tone sequences of Margi. With the sole addition of a generic NOCONTOUR constraint, the basic grammar fragment for simplex tonal spreading derives both the gaps in Margi tonal patterns (no falling contours, no high–low patterns in both underived and derived contexts) and the appearance of contour simplification.

Apparent gaps in the input may be due to constraints on the output.

Lowly ranked, robustly violated constraints may emerge to play a role in restricted contexts.

9 Conclusion

The interest in Optimality Theory derives from the possibility of accounting for a wide range of complex phonological phenomena by the interaction of a small number of simplex constraints. The claim of universality, that all constraints are universal, is of interest not because baroque conglomerate constraints are being analyzed as "universal", but because simplex constraints, often externally motivated, can be combined in various rankings to produce richly different sets of grammars. A single constraint may be highly ranked in one language, presenting a pattern that is virtually surface–true; in another language, the same constraint may play a role in a large class of environments yet be roundly violated elsewhere; and finally the same constraint may play a role only in some minor class of contexts in a third language.

In attempting to give an idea of the types of interactions that play a role in the analysis of featural phenomena, this chapter has examined different constraint families that affect the various features of which phonological segments are composed. The discussion has addressed only a small fraction of the sorts of featural behavior that are attested in natural language, and only a fraction of the sorts of patterns that have already been discussed in Optimality Theory.[12] Syntagmatic constraints have been shown to govern the acceptability of featural sequences: changes in feature values are dispreferred; articulatory inertia is highly valued. Phonetically motivated featural dependencies cause certain paradigmatic feature combinations to enhance each other, and others to repel each other. Such featural constraints define segment inventories, and their interaction can collapse featural distinctions in some contexts and create new ones in others. Individual features may be the sole markers of some morphemes, and features may serve to mark off morphological classes such as the root and word.

Within this interactive network, a variety of families of constraints impose different and conflicting requirements on phonological representations. Faithfulness constraints, governed by the push–and–pull inherent in the resolution of constraint conflict, trace a meandering path from sound to meaning, buffeted by the attempt to satisfy an array of mutually incompatible phonological demands.

No grammar can possibly have all of its words satisfy all phonological constraints. The inevitable tension is resolved in each language by assigning particular rankings to the conflicting constraints, and the range of possible rankings gives us the rich diversity that we observe in the phonological systems of the world's languages.

---

[12]See, for example, McCarthy (1995) for an application of Optimality Theory to metathesis, Jiang–King (1996) on the featural interactions between tone and vocalic features, Padgett (to appear) on implications for feature geometry, Pater (to appear) on coalescence, and Cole & Kisseberth (1994) and Pulleyblank (1994) on neutrality effects in harmony.

4

Optimality Theory and Morphology

Kevin Russell

There are many insights that can be gained by applying the ideas of Optimality Theory to the structure of words. This chapter presents some of the central ideas and questions of morphology, how linguistic theories have dealt with these ideas and questions in the past, and some of the ways that OT offers new solutions.

1 What Morphology Deals With — Morphemes

Morphology deals with the borderline between the predictable and the arbitrary in language. Many areas of language operate in a purely predictable way, for example, sentence meanings. The term **compositionality** is often used for the idea that the meaning of a whole is a function of, or composed from, the meaning of the parts. Any speaker who knows the meanings of *cats*, *likes*, and *cottage cheese*, and the rules for constructing English sentences, also knows the meaning of the whole sentence.

(4.1)   cats like cottage cheese = cats + like + cottage cheese

The phrase *cottage cheese*, on the other hand, is not perfectly compositional. A speaker who knows only the meanings of *cottage* and of *cheese* will not be able to predict the exact meaning of the entire compound, because an extra unpredictable element of meaning has found its way in. If a similar unpredictability entered into the interpretation of the whole of (4.1) and every other sentence of English, communication would become impossible. Compositionality allows human languages to construct an infinite number of meaningful sentences out of a finite number of building blocks.

Compositionality also applies within words. In most languages, words can be analyzed into smaller pieces, each contributing to the overall meaning of the word in a predictable way (although the deeper you go into the words of a language, the more likely you are to come across not–quite–predictable situations like *cottage cheese*). So, any English speaker who knows the meaning of the sequence of sounds [kæt] and knows that the [s] suffix carries the plural meaning will be able to predict the meaning of the combination [kæts].

But now we've reached the limits of this kind of analysis. We can't take the pieces apart any further. It doesn't make any sense to say that the meaning of *cat* is the sum of the meaning of [k], of [æ], and of [t]. Children learning English don't get any help from compositional rules in this case; they simply have to memorize the fact that the arbitrary string of sounds, [kæt], is associated with meaning of a furry animal that meows and scratches furniture. A **morpheme** is an association between a string of sounds and a meaning, for example, the association between [kæt] and 'cat' or between [s] and plurality. The string of sounds itself is called a **morph**.

A simple definition of morphology is the study of the way in which morphemes combine to form words. It is commonly assumed that knowing a particular language includes knowing the words of that language (its **lexicon**), and that these words have to be learned individually. It would be more accurate to say that it is the morphemes that have to be learned individually, since at least part of the process by which words are formed from morphemes is compositional.

> Morphology is the study of how morphemes combine to form words.

A number of questions for the study of morphology arise from this way of looking at things. First of all, if knowing a morpheme means knowing that a particular sequence of sounds and a particular meaning are associated, we should be able to spell out exactly what these two kinds of knowledge are and the nature of their union. For example, the knowledge of the sound pattern is presumably something more abstract than some single instance of the sound pattern—it is unlikely that I turned on a mental tape recorder the first time one of my parents said the word *cat* to me and that I've simply been replaying the tape ever since.

In morphology, we might broadly classify these detail questions into the syntax, semantics, and phonology of word structure.[1] For the syntax of word structure, it is not enough to say that morphemes can combine together. We need to explain which morphemes can combine and in what order. While the noun stem *cat* and the plural suffix *-s* can combine, *cat* and the progressive verb suffix *-ing* cannot, any more than in sentence syntax the word *the* and the verb *think* can combine to form a phrase. We need to deal with the order of the morphemes, why the plural of *cat* is [kæts] and not [skæt] or even [kæst]. For the semantics of word structure, we need to specify how the meanings of

---

[1]I am using the terms "semantics", "phonology", and "syntax" here in the most general sense: what do the pieces mean, how do you realize them physically, and how are they organized (in what linear order and what hierarchical structures)? In the study of sentences, the pieces are words. In morphology, the pieces are morphemes. (Without any further qualification, "syntax" usually refers just to the syntax of sentences.)

morphemes combine to form the meanings of whole words, and this part of the theory should also be able to deal with the not–quite–compositional cases. For the phonology of word structure, the most immediate problem is that the string of sounds associated with a certain meaning is not always the same in different words. The English plural morpheme, for example, has three different associated sound patterns or morphs: [s], [z], and [əz]. Which one will be used depends on the preceding sound of the noun stem.

(4.2)   *cats*   [kæt–s]   *dogs*   [dɑg–z]   *roses*   [roz–əz]

In describing this situation, linguists often say that [s], [z], and [əz] are **allomorphs** of the same morpheme. A theory of morphology must be able to explain how the right allomorph of a morpheme is chosen.

Although there are many interesting questions to be asked and answered about the semantics of word structure and in other areas of the syntax of word structure, in this chapter I concentrate mostly on the phonological and some of the syntactic aspects of word structure. Three of the questions I consider are given below.

---

1. What pattern of sounds is associated with each morpheme?
2. How is the right allomorph chosen for each context?
3. What determines the order of morphemes in a word?

We first consider how these questions were approached in Generative Phonology (see Chapter 1 for discussion of Generative Phonology) and then explore the different approach suggested by the leading ideas of OT.

2   The "Classical Approach" to Morpheme Form and Order

It is impossible to talk about *the* pre–OT approach to any of the questions at the end of Section 1. There has never been unanimous agreement on any question concerning morphology. Even something as basic as the existence of morphology is disputed. Do we need a separate component or module of the grammar, on a par with phonology and syntax, to deal just with the structure of words, or can all the work be done by independently needed principles of phonology, syntax, and semantics?

With questions this fundamental still unresolved, it is not surprising that a lower–level detail question like Question 3 inspires even less consensus. In syntax for example, the order of the head of a phrase with respect to its specifier and complement (see Chapter 5) has received widely divergent treatments. Some have claimed that word order is fixed by fundamental and universal properties of syntactic structure, while others maintain that it is a relatively superficial decision made by phonology. Phonologists in turn have tried to ship the problem back again, glossing over it on the assumption that linear order has already been determined by the syntax or morphology (or occasionally passing the buck in a different direction, treating linear order as a relatively superficial decision made by phonetics). There is so little agreement on where exactly the order of morphemes gets specified that I do not attempt to outline the variety of positions which have been held.

Questions 1 and 2 were much closer to having a consensus, at least among phonologists working in the generative tradition. In this section, I sketch the broad approach to these questions taken by Generative Phonology, using the English plural as an example. I then discuss some of the problems that the approach runs into.

While it might appear that the English plural morpheme has three different phonological representations associated with it, plus a set of directions for choosing the right one, Generative Phonology has tried to get as much mileage as possible from the assumption that every morpheme, including a problem case like the English plural, has exactly one **underlying representation** (UR).

Two central assumptions of Generative Phonology:
1. Every morpheme has exactly one underlying representation.
2. Any variation in the surface forms of a morpheme is attributed to the action of phonological rules.

The classical picture for the derivation of a word is this. The underlying representations of the morphemes involved are taken and strung together, or concatenated, in the right order, forming the UR for the word as a whole. Then a series of phonological rules apply to this representation—inserting, deleting, and changing certain sounds whenever the conditions are right. The pronunciation of the word is based on its **surface representation** (SR), the representation that is left after all the rules have applied to the UR. The phonological rules are responsible for creating the different allomorphs associated with a morpheme. To derive the words *cats* [kæts] and *dogs* [dɑgz], for example, the speaker takes the URs of the noun stems (/kæt/ and /dɑg/) and the UR of the plural morpheme (/z/), concatenates them to form /kæt–z/ and /dɑg–z/. The phonological rules of English then apply, including the one which changes a word–final voiced z into a voiceless s if it follows another voiceless sound, like t. (See Chapter 3 on features for "voiced/voiceless".) At the end of the derivation, the representations are something like [kʰæts] and [dɑgz].

(4.3)   A derivation in Generative Phonology

UR of morphemes in *cats*	/k æ t/ / z/
concatenation	↘ ↙
UR of the word *cats*	/k æ t – z/
phonological rule: z → s / t __	↓
intermediate representation (IR)	[k æ t – s]
other rules	↓
SR	[kʰæts]

The rule changing z to s is not just a way to sneak the other allomorphs in through the back door while still claiming that there is only one UR. The rule operates more generally in English. In fact, not just the plural, but all the single–consonant z–like morphemes at the end of a word act like this, including the singular possessive suffix usually spelled *'s*, the plural possessive suffix *s'*, the contracted *'s* form of *is*, and the contracted form of *has*. We would be missing a generalization if we were to specify separately the rules for choosing the right versions of each of these morphemes. Moving to a unique UR plus a set of general rules seems to be more in keeping with our original motivation for morphemes: the rules express those facts which are more general, predictable facts about the language as a whole, while the URs for individual morphemes contain only those unpredictable facts that are part of the arbitrary associations between meanings and sound patterns.

Three Challenges for the Classical Approach

The first major challenge for this classical way of deriving the variants of morpheme is that *the changes that phonological rules make to URs are not random*. They usually make good sense. In the English plural, the rule that changes *z* to *s* after a *t* has strong motivation. It forces the two consonants in the *t+z* cluster to agree in voicing, that is, in whether the vocal cords are vibrating, thus sparing the vocal cords from having to turn vibration on or off abruptly in the middle of the cluster. It's the kind of rule that is found in language after language. The naturalness of a rule like this was not well explained by the theory, especially in earlier versions of Generative Phonology. Usually, a rule that did the opposite, forcing the two consonants to *dis*agree in voicing, would have been just as natural in a rule–based system, despite its relative rarity in the languages of the world. To the extent that Generative Phonology allowed highly unnatural rules to be on an equal footing with widely attested rules, it failed to accurately reflect the linguistic competence of human beings.

There are also problems for the basic idea that words are composed of well–defined pieces called morphemes, no matter what kind of rules you allow, problems which have led some researchers to radically revise or even abandon the idea of morpheme (e.g., Matthews 1972a, Anderson 1992). Often there is no way to divide up a word into continuous stretches and assign a meaning to each stretch. The root–and–pattern or templatic morphologies of Semitic languages are a well–known example of this, as illustrated in the difference between the following forms of the verb meaning 'write' in Classical Arabic:

(4.4)	kutiba	'it was written'
	kattaba	'he made (someone) write'
	kuutiba	'he was corresponded with'

The phonological information associated with the meanings of the verb stem 'write', the tense, the voice (active or passive), whether the verb is simple or a causative, and so on, is all intertwined in ways that cannot be separated with a meat cleaver, the way English *cats* can. Autosegmental Phonology had a fair bit of success dealing with these cases in the 1980s, although it required interesting revisions to the idea of concatenation; morpheme combination could no longer be seen as simply laying down the URs end to end.[2] More recalcitrant problems for the idea that all morphemes are discrete sections of representations come from irregularities or subregularities. Consider the sub–class of English irregular past tense verbs illustrated by the following.

(4.5)   hide/hid, lead/led, hit/hit, put/put, buy/bought, catch/caught, ...

There's something unsatisfactory about treating these as complete irregularities on the scale of *am/was*. While we cannot cut [hɪd] into two pieces, [haɪd] and [d], it is probably no accident that the clear majority of irregular pasts end in [t] or [d] just like the regular ones do. Ideally, we would like a theory of morphology which could cope gracefully with these intermediate cases.

Perhaps the most serious shortcoming is that the only way the classical rule–based approach has to relate a UR and its surface form is by making a sequence of changes that take place in virtual, if not real, time. There are a number of cases where this model simply does not work. The ones which have attracted the greatest attention in the literature involve the phenomenon of reduplication. In the rest of this section, I discuss a reduplication process from Paamese, an Oceanic language of Vanuatu described in Crowley (1982), and I outline the problems it causes for a classical analysis.

Challenges for Generative Phonology:
1. Unnatural rules are as easy to write as natural rules.
2. Many words cannot be easily separated into morphemes.
3. Applying phonological rules sequentially often does not work.

Paamese Reduplication

Many languages mark certain meanings by copying all or part of the stem, a phenomenon known as **reduplication**. In Paamese, reduplication is used to mark a number of semantic modifications for verbs, such as habituality, randomness of action, and detransitivization. One of the three patterns for reduplication copies the first two syllables of the verb stem. (Adjacent vowels in Paamese belong to different syllables. See Chapters 1 and 2 concerning syllabification.)

(4.6) | *Simple verb* | *Reduplicated verb* | *Gloss* |
|---|---|---|
| hiteali | hite–hiteali | 'laugh' |
| hotiini | hoti–hotiini | 'find' |
| hulai | hula–hulai | 'spray' |
| saa | saa–saani | 'give' |

(The other patterns copy the last two syllables or only the first syllable.) The full or partial copy of the stem is usually called the **reduplicant**, and the stem from which it was copied is called the **base**.

---

[2] The Morpheme Tier Hypothesis is a good example of the increasingly abstract answers which Autosegmental Phonology explored for question 3 above (see, e.g., McCarthy 1989).

Marking a meaning by reduplication poses a challenge for the assumption that every morpheme has a single unique underlying representation. The English plural suffix had only three allomorphs, [s], [z], and [əz], and the small differences between them could be attributed to the actions of phonological rules. The reduplicative morpheme in Paamese, though, seems to have a different allomorph for every verb stem it attaches to: [hite], [hoti], [saa], and so forth. It is hard to imagine what its UR might be or any set of phonological rules which could result in changes this drastic.

Autosegmental Phonology had some success in making even reduplication fit the generative mold, using the idea that the UR of a reduplicant was a very underspecified representation that encoded very little else than prosodic information. (See Chapter 3 on underspecification.) The UR of the reduplicant morpheme in Paamese could be analyzed as simply a sequence of two empty syllables. The presence of empty prosodic structure triggers an automatic process which copies the segments of the base. In the ensuing game of musical chairs, as many of the copied segments as possible squeeze into the empty syllables (subject to the conditions on what counts as a legal syllable in the language), and those that cannot fit are deleted. In such a model, the derivation of the Paamese word *hite–hiteali* would look like the following.

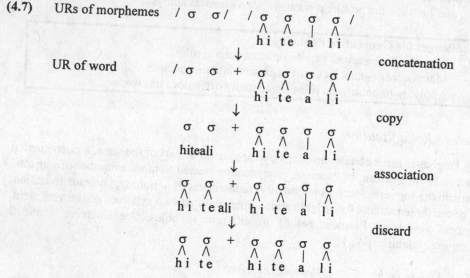

(4.7) URs of morphemes

concatenation

UR of word

copy

association

discard

This approach saves the assumption that morphemes have URs. The meanings (habituality, randomness, and so on) are marked by a stripped–down UR consisting of two syllables, not by some copying rule.

However, the "automatic copying process" turned out not be so automatic. In some languages, copying happens fairly early, with the reduplicant and the base then pursuing their separate lives through the derivation. Phonological rules that change the base do not change the reduplicant, so that the reduplicant could end up looking more like the UR than the base does. For example, in Lakhota, a Siouan language of North and South Dakota, the rule that changes an *s* to a *z* between two vowels applies after copying. By the time the *s* of the base *yas* in (4.8) changes to *z*, it is too late for there to be any effect on the reduplicant. In other languages, copying seems to occur much later, with a number of phonological rules changing the form of the base, these changes then being copied into the reduplicant even though the reduplicant might not have met the conditions for the original rule to apply. For example, the Javanese rule that deletes an *h* between two vowels applies before copying, so that the reduplicant in (4.9) ends up with no *h* either, even though its *h* would not have been between two vowels (McCarthy & Prince 1995).

(4.8) yas–a → yas–yas–a → yas–yaz–a *yaz–yaz–a

(4.9) bədah–e → bəda–e → bəda–bəda–e *bədah–bəda–e

The most troublesome cases are those where both copying and a phonological rule apply, but neither order (first copy, then apply the rule; or first apply the rule, then copy) gives the right result. Paamese has examples of such a situation. We might expect that the reduplicated form of the verb *muni* 'drink' would be *muni–muni*, simply copying the first two syllables of the base. In fact, the right form is *munu–munu*.

(4.10) muni̱ munu̱–munu̱ 'drink' *muni̱–muni̱, *munu̱–muni̱
       luhi̱ luhu̱–luhu̱ 'plant'
       uhi̱ uhu̱–uhu̱ 'blow'

Sequences with *i* and *u* on opposite sides of a morpheme boundary are avoided in Paamese. This is a fairly strong tendency which can be seen operating in places other than reduplication. We could formulate this as a classical rule as in (4.11).

(4.11) i → u / __ – (C) u

This rule interacts with the copying process of reduplication. However, neither ordering of copying and rule (4.11) produces the correct surface forms in (4.10). If you try to copy first and then apply rule (4.11) to change the vowel, you end up with the incorrect [luhu–luhi].

(4.12) / σσ / – / luhi /
       luhi – luhi copy
       luhu – luhi rule (4.11)

If you do it the other way around, applying rule (4.11) first and then copy, when you try to apply rule (4.11) its environment is not met. The representation has no *i* followed by a morpheme boundary followed by an *u*, so it changes nothing. After copying you end up with [luhi–luhi], which is also the wrong answer.

(4.13) / σσ / – / luhi /
               luhi rule (4.11), nothing to apply to
       luhi – luhi copy

This is a simple example of an **ordering paradox**: the environment for changing an /i/ to an [u] isn't there until after you copy the base, but by then it's too late to make the change and have it automatically affect both the base and the reduplicant. What seem to be going on is that the base and the reduplicant keep some kind of association with each other even after the copying is finished. Anything you do to one of them automatically affects the other one. We might try to capture this with another rule. First do the copying, then apply rule (4.11) to change the final vowel of the reduplicant, resulting in [luhu–luhi], and then have some other process come along, notice the change to the reduplicant, and transmit it back to the base.

(4.14)　　/ σσ / 　–　　/ luhi /
　　　　　luhi 　–　　luhi　　　copy
　　　　　luhu 　–　　luhi　　　rule (4.11)
　　　　　luhu 　–　　luhu　　　mysterious retransmission process

The classical approach to phonology and morphology does not have any elegant and convincing way of doing this. The difficulty here is caused by the theory's requirement that the copying operation be performed at a particular point in the derivation, either before or after other phonological rules. The data suggest, though, that the similarity between the reduplicant and the base is something that Paamese actively seeks to maintain and is more than just an accident of a derivation's history.

> The similarity between a reduplicant and its base is an important and continuing relationship, not the side effect of a copying process.

In discussing the problems facing these central assumptions of the classical approach, I may have given the impression that they are all insuperable. (For a survey of the many ingenious attempts to overcome them by revisions to the framework like Autosegmental Phonology, see Kenstowicz 1994). But the band–aids that were applied, such as enriched representations and principled general constraints, were needed so consistently in the same places that it raised suspicions over whether the ideas of phonological rewrite rules and sequential derivations might not be the fundamental problems. One might view Optimality Theory as the band–aids getting together, realizing their own power, and deciding that they could get along quite nicely without the patient.

3 Applying OT Ideas to Morphology

Problems with reduplication, like the Paamese example just presented, are among the main motivations for the development of Optimality Theory. The problems with a classical analysis of Paamese arise when we try to account for the similarity between the reduplicant and the base as the result of a copying operation that occurs at a particular point in a derivation. A theory like OT, with no processes or derivations in time, is ideally suited for a better analysis.

In this section, I outline a simple OT analysis of Paamese reduplication. Then we see how some of the central ideas of this analysis can be extended to a general theory of constraints that enforce similarity between two representations and the implications that this theory has for OT's answer to Questions 1 (how to encode the phonological information associated with a morpheme) and Question 2 (how to determine the right allomorph in each situation). Then we turn to Question 3 (how to get the morphemes in the right order).

OT Meets Paamese

Rather than treating the similarity between the reduplicant and the base as the side–effect of a copying process, OT treats it as the result of a constraint, or a set of constraints, that enforces that similarity. We can informally describe this constraint as:

(4.15) RED=BASE: The reduplicant (the copy) and the base (the original) should be identical.

Constraints of this type, which enforce similarity between two types of representations—a reduplicant and a base, or an underlying representation and a surface representation—are often referred to as **faithfulness** constraints. Like any other constraint in OT, there is no guarantee that a faithfulness constraint will always get its way. It may be outranked by other constraints which can force violations by removing all the candidates which would have satisfied the faithfulness constraint before it ever has a chance to see them. For reduplication, this can result in only partial copying (see Chapter 1 for further discussion of faithfulness constraints.)

One of the choices many languages use to force partial reduplication is a constraint that the reduplicant have a certain prosodic form, that it consist of a certain kind of syllable or foot. (See Chapter 2 on the notion of "foot" and "heavy syllable".)

(4.16) **Some possible prosodic conditions on the reduplicant:**
Reduplicant = heavy syllable
Reduplicant = iambic foot
Reduplicant = minimal word

The Paamese pattern that we've been looking at requires that the reduplicant be a foot consisting of two syllables. The correct reduplication for 'laugh' is *hite–hiteali*, even though *hiteali–hiteali* would satisfy RED=BASE even better. I do not discuss RED=BASE in Paamese any further, since the words we are most interested in have only two syllables in the first place and the prosodic requirement forces no violation of faithfulness for them.

As we have already seen, Paamese avoids having a syllable containing *i* followed by a syllable containing *u* on the other side of a morpheme boundary. It is quite possible that this avoidance could be broken down into a handful of properly ranked well–motivated universal constraints—it is fairly common, for example, for a language to require neighboring vowels which are both high to also agree in whether they are front or back (see Chapter 3). But the details of this are not very relevant for the discussion, so I simply assume that Paamese has a language–particular constraint against the undesired situation:

(4.17) *i+u: Sequences of *i* and *u* across a morpheme boundary are not allowed.

Unsurprisingly in an OT framework, this constraint doesn't always get its way, but we can see its effects in reduplication.

The last piece of the analysis is a second kind of faithfulness constraint, this time between the base or input and the *underlying representation* of the verb morpheme:

(4.18)   BASE=INPUT: The base and its input should be identical.

Together, the three (sets of) constraints, RED=BASE, BASE=INPUT, and *i+u, can result in several different patterns of reduplication, depending on how they are ranked.

Given the input /luhi/ and the instruction to produce a reduplicated form, GEN produces an infinite number of candidates. (See Chapter 1 on GEN.) Most of these bear so little resemblance to the winning candidate or to the UR, and violate faithfulness constraints so flagrantly, that we don't need to consider them seriously, e.g., [kæt], [hiteali–luhi], [ihul–mango], [e]. Some of them are closer, e.g., [luh–luhi], but still violate the constraints just proposed in a fairly obvious way. The three candidates which are the most interesting for the present analysis are the following, each of which violates one of the constraints in (4.15), (4.17), and (4.18), while obeying the other two.

(4.19)   *reduplicant* base

luh**i** – luh**i**	violates	*i+u
luh**u** – luh**i**	violates	RED=BASE
luh**u** – luh**u**	violates	BASE=INPUT

Each of these three candidates is the winner in a constraint hierarchy where the constraintt it violates is the lowest ranked. The winner is *luhi–luhi* if *i+u is ranked lower than both RED=BASE or BASE=INPUT.

(4.20)   **A ranking in which RED=BASE, BASE=INPUT » *i+u**

UR: /luhi/	RED=BASE	BASE=INPUT	*i+u
☞ *luhi*–luhi			*
luhu–luhi	*!		
luhu–luhu		*!	

In (4.20), the winner has a dispreferred sequence of *i* and *u* syllables, but this violation of *i+u is forced because the ranking in (4.20) places more value on the reduplicant being identical to the base and the base being identical to its UR.

The winner is *luhu–luhi* if RED=BASE is the lowest ranked of the three.

(4.21)   **A ranking in which BASE=INPUT, *i+u » RED=BASE**

UR: /luhi/	BASE=INPUT	*i+u	RED=BASE
luhi–luhi		*!	
☞ *luhu*–luhi			*
luhu–luhu	*!		

Both (4.20) and (4.21) are possible rankings, and both can be found in languages where the demands of faithfulness and a language–particular constraint like *i+u conflict. They give the same results as the classical analyses that ordered the i→u rule before copying as in (4.13) and after copying as in (4.12) respectively. The Lakhota example given earlier would use a ranking like the one in (4.13) and the Javanese example a ranking like (4.12). The ranking Paamese actually uses, however, is one where BASE=INPUT is the lowest ranked, and this gives the result which the classical model could not get simply by ordering copying and the i→u rule.

(4.22)   **The ranking Paamese uses: *i+u, RED=BASE » BASE=INPUT**

	*i+u	RED=BASE	BASE=INPUT
luhi–luhi	*!		
luhu–luhi		*!	
☞ *luhu*–luhu			*

There is no ordering paradox in the OT treatment of Paamese. The fact that the final vowel of the base seems to be affected by a rule (despite the fact that it doesn't meet the environment for the rule) follows naturally from the constraints of the language: the final vowel of the reduplicant is forced to be a *u* and the final vowel of the base is forced to be the same as the final vowel of the reduplicant. The constraint which could have kept the base's final vowel as an *i*, BASE=INPUT, is ranked too low to have an effect on this. By the time the candidate set reaches BASE=INPUT, all the candidates with [luhi] as the base have been removed.[3]

---

[3]Some more details on Paamese: Crowley discusses three patterns of reduplication, a one–syllable prefixed reduplicant, a two–syllable prefixed (the pattern we have been discussing), and a two–syllable suffix. Different verbs select different patterns; many select more than one pattern, using the distinction in pattern to differentiate between some of the semantic values of reduplication (e.g., detransitive, habitual). Crowley offers interesting, but not clearly compelling, reasons to place *luhu–luhu* in the two–syllable prefix class rather than the two–syllable suffix class. (If the reduplicant in *luhu–luhu* were indeed a suffix, our OT analysis would change somewhat, but it would still remain as a problem for a classical analysis: the *i* of the stem (the first half) could only change to *u* after the copying operation has provided the environment, but by then it's too late for the new vowel to be copied to the reduplicant.) Crowley notes that the i→u change does not affect reduplications which are clearly in the two–syllable suffix class: *tukuli* 'creep' reduplicates as *tukuli–kuli*, not *tukulu–kulu* nor *tukulu–kuli*. Accepting Crowley's assignment of *luhu–luhi* etc. to the two–syllable prefix class, it's easy to account for the difference: faithfulness between a two–syllable prefix reduplicant and its base (our RED=BASE) outranks faithfulness between the base and its UR (our BASE=INPUT) which outranks faithfulness between a two–syllable *suffix* reduplicant and its base.

In addition, word–final vowels are usually subject to deletion. The reduplicated form of *luhu–luhu* usually surfaces as [luhu–luh] unless it is followed by some suffix. If Crowley's assignment of these forms to the two–syllable prefix class is correct, then this is a case where the reduplicant is more faithful to the UR of the stem than it is to the base, and more faithful than the base is to the UR, a situation which McCarthy & Prince (1995) claim is impossible.

Selecting the Correct Allomorph: Faithfulness and Correspondence

Faithfulness. In tableau (4.22) in the analysis of Paamese reduplication, the constraint BASE=INPUT does not seem to be doing a lot of work. It is, however, a very important constraint in the phonology of Paamese, as it is in most languages. In Paamese, faithfulness between base and UR is ranked lower than faithfulness between reduplicant and base (RED=BASE) and the language–particular constraint *i+u, but it does rank higher than a number of other constraints. Tableau (4.22) shows only the competition between three candidates—for the purposes of printing the tableau in a small space, I left out all the other candidates with the promise that they would all fare worse than the winner. While BASE=INPUT doesn't seem to be doing anything interesting in (4.22), it is one of the constraints that does the most work in ruling out the hordes of other candidates. (See Chapter 1, Section 5 on this same point.)

Specifically, the candidate set has not only the candidates in (4.19) and junk like [kæt] and [blag]. It also has a number of well–formed Paamese words which are optimal candidates in some circumstances but which happen not to mean 'plant'. Some of the also–rans are the attested reduplications [hite–hiteali], [saa–saani], and [munu–munu]. These satisfy as optimally as possible the higher–ranking constraints RED=BASE, *i+u, and the constraint that the reduplicant have exactly two syllables (which overrules even RED=BASE). They will still survive in the candidate set after all these higher–ranked constraints have applied. It is BASE=INPUT which decides the case—the candidates may be fine words of Paamese, but not as realizations of the verb whose UR was the input to GEN. In (4.23), violation marks in the BASE=INPUT column indicate the number of segments of the UR which are not present in the base.

(4.23)   **How /luhi/→[hite–hiteali] loses**

UR: /luhi/	*i+u	RED=BASE	BASE=INPUT
☞ [luhu]–[luhu]			*
[hite]–[hiteali]			**!**
[saa]–[saani]			**!**
[lu]–[lu]			**!
[kæt]	*!		****

This example illustrates the dominant way of thinking about morphemes and their realizations in OT, and the ideas serve as OT's answer to Question 1 on page 104. In OT, the phonological information associated with a morpheme is still assumed to take the form of an underlying representation, just as it did in Generative Phonology, and it is still assumed that each morpheme should have a single UR. But there is a difference in how this UR relates to the surface form. In Generative Phonology, the surface form is the battle–scarred carcass of the UR, having been cut, inserted into, and otherwise mutilated by a battery of phonological rules. Beyond that fact, similarities between the two are coincidental. A piece of a surface form that is identical to its UR is simply a piece which has managed to survive unscathed while nasty things were happening to the pieces on either side of it. There is seldom a satisfying explanation for why a particular piece is spared while its neighbors are not.

> **OT's answer to Question 1 on page 104:**
> As in Generative Phonology, the sound pattern of a morpheme is encoded in an underlying representation.

OT has a different conception of the relationship between the UR and the surface form, that is, the winning candidate. The two come from different places: the UR from the lexicon, the surface form from the infinite candidate set produced by GEN. The UR does not turn into the surface form. (We often talk about the candidate set as being created from the UR by an infinite number of deformations. But since GEN will, in theory at least, produce exactly the same infinite set of deformed candidates no matter what its input is, we might as well regard the candidate set as a primitive entity in its own right.) Any similarities between the UR and the surface form are the result of constraints that actively require that similarity. Candidates which are different from the UR will collect violation marks in EVAL, which will usually result in the sudden death of those candidates. Similarities between URs and surface forms are the responsibility of faithfulness constraints; differences are the result of faithfulness constraints being outranked.

> **OT's answer to Question 2 on page 104:**
> Rankings between faithfulness constraints and other phonological constraints determine the selection of the winning allomorph.

One way of looking at this is to see the UR as a kind of prototype of the morpheme, an ideal that all the various allomorphs of the morpheme aspire to. The main difference is that OT is not a prototype theory, it is a classical categorization theory that uses necessary and sufficient conditions to determine what is good and what is not. These necessary and sufficient conditions are given by the operation of the constraint hierarchy subject to strict dominance.

In the Paamese analysis, we count segmental mismatches between the reduplicant and either the base or the input in order to evaluate the success of the constraints RED=BASE and BASE=INPUT. However, there are other way that the violations might be counted for comparison against each other. If perfect faithfulness must be violated, languages show definite opinions on which violations are better than others, and these preferences can operate on a scale as small as single features. In Paamese, *luhi–luhi* is ruled out by *i+u, but to avoid this violation, *lihi–lihi* does just as well as *luhu–luhu*. In Paamese it is more important to be faithful to an underlying back vowel than to an underlying front vowel. A good theory should allow faithfulness constraints to express these preferences rather than imposing a monolithic requirement of complete identity. (See Chapter 3 for discussion.)

Correspondence. A more complete approach to faithfulness constraints is given by the sub–theory of OT known as **Correspondence Theory** (McCarthy & Prince 1995). Correspondence constraints can hold between any two representations, for example, between a UR and a candidate or a part of a candidate (such as BASE=INPUT in the Paamese example), between two parts of a candidate (such as base and reduplicant in Paamese RED=BASE), or even between a candidate and another well–formed word of the language. We'll look more at this last possibility later.

Correspondence constraints rely heavily on correspondence relations between two representations. They ask questions like: Does this segment in the candidate have a segment in the UR that corresponds to it? Do the corresponding segments have the same values for the feature [voiced]? Do two pairs of corresponding segments have the same order in both representations?

GEN sets up the correspondence relations that these questions refer to. For example, given an input consisting of the verb stem /hiteali/ and the reduplicative morpheme, GEN creates each possible set of correspondence relations between the UR /hiteali/ and each possible phonological representation, and also each possible reduplication–type correspondence relation between the pieces of each possible representation. Strictly speaking, a candidate should be seen not as a representation, but as the combination of a possible surface representation, the associated UR, and one or more correspondence relations among the parts. The winning candidate for the reduplicative form of /hiteali/ is shown in (4.24), where numeric subscripts represent the correspondence relation GEN has set up between the UR and the surface form, and letter subscripts represent the correspondence relation between the reduplicant and the base:

(4.24)   UR:

                                 / h i t e a l i /
                                   1 2 3 4 5 6 7
                                   | | | | | | |
                                   1 2 3 4 5 6 7
         SR: [h i t e]   [h i t e a l i]
                 a b c d   a b c d

The correspondence relations that GEN sets up need not be one–to–one, or for that matter even sensible. To illustrate, five candidates for English *cat*, are given in (4.25), complete with the UR, an SR, and correspondence relations ranging from reasonable to wild.

(4.25)

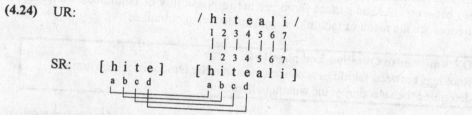

Correspondence constraints come in families that look at particular aspects of the correspondence relations. Some of these families are listed in (4.26). Most of these are phrased in terms of correspondence relations between surface form and UR, though families of correspondence constraints can refer to any other kind of correspondence relations as well.

(4.26)   **Some correspondence constraint families**

MAX   Every segment in the UR should correspond to a segment in the surface form (cf. FAITH in Chapter 1).

DEP   Every segment in the surface form should correspond to a segment in the UR (cf. FAITH in Chapter 1).

IDENT(F)   Corresponding segments should have identical values for feature F.

LINEARITY Corresponding segments should be in the same order in both representations.

CONTIGUITY If two segments are adjacent in the UR, the corresponding segments should be adjacent in the surface form, and vice versa.

ANCHORING If a segment is at one edge of the UR, the corresponding segment should be at the same edge of the surface form, and vice versa.

The families of constraints on a correspondence relation can be ranked at different places in the hierarchy: MAX for the reduplicant–base correspondence might be ranked more highly than DEP. The same family for different correspondence relations can also be ranked at different places. In Paamese, IDENT ([back]) for the reduplicant–base relation (part of what we called RED=BASE) is ranked more highly than IDENT([back]) for the base–UR relation (part of what we called BASE=INPUT).[4]

Morpheme Order and Generalized Alignment

The first part of the Paamese word *hite–hiteali* has only two syllables and is clearly the reduplicant rather than the base. But, if GEN is free to do any kind of reanalysis at all, then in addition to the candidate and correspondence relations in (4.24), there should also be a candidate with the following correspondence relations:

(4.27) UR:

SR:

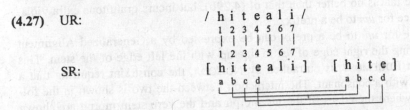

Some constraint will have to weed out (4.27) in favor of (4.24). This constraint can be informally stated as: the reduplicant is a prefix rather than a suffix.

Generalized Alignment. OT has a subtheory called **Generalized Alignment** which offers a more rigorous way to specify the constraints that are responsible for morpheme order, as well as many other kinds of patterns (McCarthy & Prince 1994). Generalized Alignment offers a way to require that an edge of one constituent coincide with an edge

---

[4]Correspondence Theory is not the only possible approach to Question I within the overall framework of Optimality Theory. Hammond (1995), Russell (1995) and Golston (1996), for example, argue that the phonological information of morphemes can be encoded directly in constraints on surface forms, rather than in constraints that compare a surface form to some other abstract representation. Golston (1996) proposes that the phonological information of a language's morphemes can be encoded in the patterns of the violations that they cause for universal constraints. These proposals take a more radical approach to the nature of a speaker's knowledge of morphemes, but they share with Correspondence Theory a rejection of Generative Phonology's central hypothesis that a speaker's knowledge of a morpheme actually *becomes* a pronounceable surface form after having survived a series of phonological rules.

of another. Saying that the reduplicant comes before the base is saying that the right edge of the reduplicant coincides with the left edge of the base.

Generalized Alignment constraints can be violated, though as with any other OT constraint the violation is as minimal as possible. A good example is provided by infixation in Tagalog. The subject–focus affix *um* is often a prefix, but often ends up located somewhere inside the stem it attaches to, rather than staying considerately right at the beginning.

(4.28)   **Tagalog *um* as a prefix**
             aral   um–aral

(4.29)   **Tagalog *um* as an infix**
             a. sulat   s–um–ulat
             b. *um–sulat
             c. *sul–um–at

We can see the preference for *um* to be a prefix when it occurs with a stem beginning with a vowel, as in (4.28). But this preference can be overruled by the syllable structure constraints of Tagalog: (4.29a) has a more optimal syllable structure than (4.29b) and is preferred even at the expense of not allowing *um* to be a prefix. But even so, *um* will be as close to the left edge as it can; placing *um* further from the left in (4.29c) results in a syllable structure that is no better than that of (4.29b), but incurs gratuitous extra violations of preference for *um* to be a prefix.

The preference for *um* to be a prefix can be expressed by a Generalized Alignment constraint requiring the right edge of *um* to line up with the left edge of the stem. This constraint, call it EDGEMOST, is outranked by NOCODA, the constraint requiring that a syllable not end with a consonant. The interaction between the two is shown in the following tableau, where the edges of the *um* morph and the verb stem morph are shown with square brackets, and syllable boundaries with dots.

(4.30)   **Tableaux for Tagalog *um* as an infix and as a prefix**

URs: /um/ /sulat/	NOCODA	EDGEMOST
.[um].[su.lat].	**!	
☞ .[s[u.m]u.lat].	*	***
.[su.l[u.m]at].	*	****!*
URs: /um/ /aral/	NOCODA	EDGEMOST
☞ .[u.m][a.ral].	*	
.[a.[um].ral].	**!	***
.[a.r[u.m]al].	*	***!*

Both the infixed version *s–um–ulat* and *sul–um–at* escape NOCODA equally well, getting one unavoidable violation mark for the final *t*, while the prefixed version *um–sulat* receives an extra fatal violation mark for the syllable ending with *m*. The candidate that EDGEMOST would have liked best is no longer in the running, but EDGEMOST *can* choose from among the remaining candidates, selecting the one where the right edge of *um* is as close to the left edge of the stem as possible: in *s–um–ulat* it is three segments away, in *sul–um–at* it is five. With vowel–initial stems like *aral*, neither the prefixed nor the infixed versions will violate NoCODA, except for the unavoidable mark given to the final *l*, so the decision is passed along to EDGEMOST, which picks the prefixed version.

Concatenation of morphemes is no more complicated than this. We do not need to assume some other mysterious component of the grammar which is responsible for feeding GEN the URs of its morphemes already strung together in the right order. Infixation in Tagalog shows that stringing morphemes together outside of phonology is not even feasible. Alignment constraints take care of concatenation naturally. We can require that the English plural marker be a suffix with an alignment constraint requiring that its left edge line up with the right edge of the noun stem. The only significant difference from Tagalog is that English has no outranking phonological constraints which are powerful enough to force the *–s* suffix inside the noun stem in violation of the alignment constraint.

OT's answer to Question 3 on page 104:
The order of morphemes is determined by Generalized Alignment constraints.

The sub–theory of Generalized Alignment proposes that there is a constraint schema, i.e., a general pattern that languages can use to create edge–alignment constraints for the categories that they are interested in. The schema can be specified for two categories and the edges that should line up.

(4.31)   **Generalized Alignment schema:**
ALIGN (Category$_1$, Edge$_1$; Category$_2$, Edge$_2$)
> where Category$_1$ and Category$_2$ are either grammatical or prosodic categories, and Edge$_1$ and Edge$_2$ can be either Left or Right

Interpretation:
> for all instances of Category$_1$, there exists some instance of Category$_2$, and Edge$_1$ of Category$_1$ lines up with Edge$_2$ of Category$_2$

The Tagalog prefixing constraint we called EDGEMOST is one possible way of specifying the parameters in (4.32).

(4.32)   **Tagalog EDGEMOST:** ALIGN (*um*, Right; Stem, Left)

That is, for every instance of the grammatical category represented by *um* there must be a stem, and the right edge of the *um* morph must line up with the left edge of the stem morph. Violations of constraints specified with this schema can be measured in terms of the number of segments that occur between the two edges.

ALIGN is not a constraint, it is a schema for creating constraints.

ALIGN is not a constraint, it is a schema for creating constraints. One important consequence of this which has not been thoroughly examined within OT is that the constraints which appear in a language's hierarchy are not necessarily universal. It would make little sense to say that constraint (4.32) of Tagalog is also present in the grammar of English (just ranked so low that we never see its effects). Mohawk has an agreement prefix for direct objects of dual number and feminine/zoic gender, which alignment constraints can refer to; it would make little sense to say these Mohawk constraints are also in the grammar of Tagalog. Of course, there are still a number of constraints which presumably form part of Universal Grammar (e.g., syllable structure constraints like ONSET and NOCODA, discussed in Chapters 1 and 2, and grounding constraints, discussed in Chapter 3), but interspersed among these are other constraints which refer to categories which only that one language is interested in.

Generalized Alignment can handle more than just concatenation. The schema can be filled in with either grammatical or prosodic categories. The concatenation constraints we've been looking at involve grammatical categories only, but uses can be found for all the possible combinations.

(4.33)   Examples of what Generalized Alignment can do:

For all		there is some		
	a. grammatical categories		a. grammatical category:	a. concatenation
	b. prosodic categories		b. prosodic category:	b. prosodic phonology
	c. grammatical categories		c. prosodic category:	c. prosodic morphology
	d. prosodic categories		d. grammatical category:	d. parsing strategies

An alignment constraint might require that the right edge of every prosodic word coincide with the right edge of a syllable, which would disallow extrametrical consonants if not overruled, or with the right edge of a trochaic foot, which would stress the second–last syllable. These kinds of prosodic–prosodic alignments result in effects which are not primarily morphological.

A language can also require an edge of a grammatical category to line up with an edge of a prosodic category. Examples of prosodic requirements on grammatical categories are numerous. We've seen that the most common Paamese reduplication pattern requires the reduplicant to consist of a single foot with two syllables.[5] Broken plurals in Arabic regularly begin with an **iambic** foot, a weak syllable followed by a strong syllable (Hammond 1988, McCarthy & Prince 1990). Hammond (1995) argues that complexities in the stress system of Spanish can be best accounted for by constraints aligning individual morphemes with prosodic feet. This class of alignment constraints can also be responsible for prosodic subcategorization, cases where an affix is fussy about the prosodic shape of the stem it attaches to. The English comparative –er requires its stem to be **trochaic**, a single syllable or strong–weak sequence of two syllables, but not a weak–strong iambic

---

[5]Strictly speaking, we need more than straightforward Generalized Alignment for this. We could require the left edge of the reduplicant to line up with the left edge of a foot, and the right edge of the reduplicant with the right edge of a foot, but we would also need a way to demand that it be the *same* foot in both cases, so that [(hi.te)] is a legal reduplicant but [(hi.te)(a.li)] isn't.

foot: (sílli)er, *(insáne)er. This requirement could be enforced in part by the constraint ALIGN (*–er*, Left; Trochee, Right).

The last kind of alignment constraint, requiring every instance of a prosodic category to line up with some grammatical category, does less work in the grammars of languages. McCarthy & Prince (1994) suggest only a rather subtle stress effect as a possible job. But this kind might be useful in making utterances easier for a listener to analyze. To the extent that Align (Stem, Left; Prosodic Word, Left) and Align (Prosodic Word, Left; Stem, Left) are both true, a listener can confidently begin a morphological and syntactic analysis of an utterance based on the prosodic clues for where stems begin. If for a grammatical and a prosodic category, there are constraints aligning both edges in both directions, then there is effectively a constraint that there be a one–to–one relation between the two:

(4.34)   a. ALIGN (Morpheme, Left; Syllable, Left)
         b. ALIGN (Morpheme, Right; Syllable, Right)
         c. ALIGN (Syllable, Left; Morpheme, Left)
         d. ALIGN (Syllable, Right; Morpheme, Right)

The set of constraints in (4.34), if unviolated, will ensure a one–to–one relationship between morphemes and syllables. Generally, languages don't behave this nicely, though Moira Yip has suggested to me that the generalization is true of Cantonese, and there are parts of other languages where it also holds, e.g., in the verb prefix system of Shona (Myers 1990). In addition, initially assuming that there is a one–to–one correspondence between morphemes and syllables may be one of the strategies children use when trying to figure out the morphological systems of their language (cf. Peters & Menn 1993).

4 Examples from English Morphology

We have seen in a general way how OT addresses the three questions for morphological theory posed at the beginning of the chapter. In this section I show how the various proposals outlined so far can fit together in an account for a relatively complex morphological problem—the behavior of the English plural, possessive, and plural possessive suffixes. I also suggest ways in which the same ideas can be applied to subregularities in English past–tense formation.

English Regular Plurals

The regular plural suffix of English shows up in three different versions, depending on the segment at the end of the preceding noun stem. If the noun ends in a sibilant consonant (s, z, š, ž, ǰ, or č), then the form is [əz], as in *roses*. If it ends in a non–sibilant but voiceless consonant, the form is [s]. Otherwise, the form is [z]. (Whether a sound is a sibilant or is voiced or voiceless is formally expressed by the features representing that sound. See Chapter 3 for discussion.)

(4.35) a. cats $[kæt]_{stem}$ $[s]_{PL}$
        b. dogs $[dɑg]_{stem}$ $[z]_{PL}$
        c. roses $[roz]_{stem}$ $[əz]_{PL}$

In Generative Phonology, this alternation is handled by choosing one of the allomorphs as the underlying representation, say /z/, and writing rule to change the UR into the other allomorphs in the right environments, as in (4.3).

In OT, the alternation can be dealt with by means of universally well–motivated constraints. It is awkward to pronounce two sibilant consonants next to each other, as you would have to in *matchs [mæčs]; using the [əz] allomorph is a way of keeping the two sibilants apart. It is awkward to switch from voiceless to voiced or vice versa in the middle of a consonant cluster; choosing the allomorph, [s] or [z], which agrees in voicing with the preceding consonant avoids this situation. Both these constraints might be seen as sub–cases of what phonologists call the Obligatory Contour Principle, which prohibits adjacent identical elements. We can specify these two constraints as follows.

(4.36) *SIB–SIB: Two sibilants cannot be adjacent.

(4.37) VOICING: Two consonants in a cluster must agree in voicing.

We also need a Generalized Alignment constraint to give the proper concatenation of the plural suffix and the noun stem:

(4.38) PL–AFTER–N: ALIGN (Plural, Left; Noun–stem, Right)

That is, the left edge of every plural morpheme must coincide with the right edge of a noun stem.

Finally, we need a constraint which penalizes the allomorph [əz]. [kæts] and [kætəz] are both conceivable ways of avoiding the VOICING violation of [kætz]. Separating the disagreeing consonants with a schwa serves the purpose just as well as changing one of the consonants does. We want to disallow the [əz] option for run–of–the–mill voicing disagreements and keep it as a last resort for sibilant disagreements. [əz] is a sub–optimal realization of the UR /z/ by several faithfulness constraints. For concreteness, let's use the following.

(4.39) LEFT–ANCHOR_plural: The leftmost segment of the plural morph corresponds to the leftmost segment of its UR.

In short, the first segment of the plural suffix should be [z]. I do not discuss how the constraint VOICING interacts with faithfulness constraints.

Some of the more promising candidates for the plural of rose are shown in (4.40). In each case, correspondence relations with the underlying representations are indicated with subscripts. The edges of the allomorphs are indicated with square brackets.

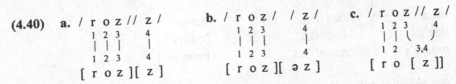

(4.40)   **a.** / r o z // z /   **b.** / r o z / / z /   **c.** / r o z // z /

Candidate (4.40a) is simple and direct. Both morphemes correspond perfectly to their URs and they are concatenated as simply as possible. Candidate (4.40b) also has simple concatenation, and satisfies PL–AFTER–N perfectly, though the plural morpheme violates LEFT–ANCHOR$_{plural}$. Candidate (4.40c) might seem an odd contender at first glance. It avoids having two sibilants in a row, but at the cost of not having the right edge of the stem line up with the left edge of the suffix. The fact that a single consonant in the candidate corresponds to two different consonants in the URs does not disqualify it. In fact I argue shortly that this is exactly what's going on with the plural possessive.

In order to choose (4.40b) as the winning candidate, our three constraints must be ranked as in (4.41). This gives the tableau in (4.42).

(4.41)   Constraint ranking:
PL–AFTER–N, *SIB–SIB » LEFT–ANCHOR$_{plural}$

(4.42)   Tableau for *roses*

/roz, z/	*SIB–SIB	PL–AFTER–N	LEFT–ANCHOR$_{plural}$
[r_1 o_2 z_3] [z_4]	*!		
[r_1 o_2 [$z_{3,4}$]]		*!	
☞ [r_1 o_2 z_3] [ə z_4]			*

English Plural Possessives

There are a number of puzzling cases in the languages of the world where it seems that more than one morpheme is occupying exactly the same space in a phonological representation. An extreme example is the root–and–pattern morphology characteristic of Semitic languages, illustrated in (4.4). Much of the effort in Autosegmental Phonology was spent on trying to make the morphemes in systems like this look as if they really could be cut apart cleanly. Less extreme but more intractable examples can be found in the irregular word forms of many languages. For example, the English plural *mice* cannot be taken apart cleanly into a morpheme meaning 'mouse' and a morpheme meaning 'plural'. For the rest of this section, I'd like to show how the ideas of OT morphology that we've seen so far can handle a simple yet subtle example where two morphemes seem to be occupying the same place, namely the English plural possessive. (For another OT analysis of English plural possessives, see Yip 1995.)

Generally, the possessive suffix of English behaves like the plural suffix in the ways we just saw. It has three allomorphs, [z], [s], and [əz], whose distribution follows the same generalizations as for the plural:

(4.43)   cat's   /kæt/+/z/_{POSS}   [kæts]
         dog's   /dɑg/+/z/_{POSS}   [dɑgz]
         mouse's   /maws/+/z/_{POSS}   [mawsəz]

Given this behavior, we might expect the *plural* possessive to have both the plural suffix and the possessive suffix, with an extra schwa to avoid having two adjacent sibilants.

(4.44)   **Expected:** /kæt/ + /z/_{PL} + /z/_{POSS} → *cats's [kætsəz]

The actual form shows no difference from either the plural or the possessive:

(4.45)
	non–possessive		possessive	
singular	*cat*	[kæt]	*cat's*	[kæts]
plural	*cats*	[kæts]	*cats'*	[kæts]

We might say that the marker that is spelled –*s'* in English orthography is just a small corner of irregularity, not two morphemes, but just a single plural–possessive suffix. The fact that it does not resemble a sequence of the plural and possessive–singular suffixes would be no more significant than the fact that the Latin genitive plural –*arum* does not look like a sequence of the nominative plural –*ae* and genitive singular –*ae*. But there are ways in which the English plural possessive is distinctly unlike the Latin genitive plural, ways which suggest that treating –*s'* as a single irregular suffix is not the most appropriate analysis. First of all, the possessive marker –*'s* is not a suffix on the noun stem, except accidentally. Rather, it is a **clitic** that attaches to the entire noun phrase representing the possessor. If this noun phrase contains a relative clause, then –*'s* will attach to the last word of the relative clause, regardless of the category of that word.

(4.46)   the guy<u>'s</u> car
         the guy I met<u>'s</u> car
         the guy I met yesterday<u>'s</u> car
         the guy I was talking to<u>'s</u> car

In contrast, the plural suffix always attaches to the stem of the head noun. (4.47a) can only be a plural, (4.47b) can only be a possessive, (4.47c) can only be both:

(4.47)   a. the guy<u>s</u> I was talking to
         b. the guy I was talking to<u>'s</u> (car)
         c. the guy<u>s</u> I was talking to<u>'s</u> (cars)

Example (4.47c) shows that the –*s* marking plurality and the –*'s* marking possession can occur independently in the same noun phrase. Their strange collapsing together into a single –*s'* occurs only when the head noun happens to fall at the extreme right edge of its phrase, as it does when the noun is not followed by any modifiers.

It is hard to come up with a convincing solution to this in a purely classical theory. (See Stemberger 1981, who recognizes that –*s'* somehow expresses both morphemes at the same time and discusses some possible analyses.) One possibility would be to let the morphological component assemble the string of abstract morphemes Stem + Plural + Possessive, and then delete either Plural or Possessive. Or we could spell out the morphemes to /kæt+z+z/ then delete one of the zs. But the classical framework forces us to choose which morpheme or which z to delete, and there are no principled grounds for choosing one over the other. Rather than deleting a morpheme, another possibility would be to "fail to insert" it, under certain defined circumstances. Again, we would be forced to make an arbitrary decision as to which morpheme should fail to be inserted. Neither approach seems to express the intuition that both suffixes are somehow present in the surface form.

With Correspondence Theory, we *can* express the intuition that both suffixes are still there. Candidate (4.40c) failed as a representation for *roses*. Nonetheless, candidates where two UR consonants correspond to the same surface segment must be considered. Using this possibility, the following seems like a reasonable representation for *cats'*, with correspondence relations to the URs as indicated.

(4.48)   The successful candidate for *cats'*

$$/ \; k \; æ \; t \; // \; z \; // \; z \; /$$
$$ 1 \; 2 \; 3 \; \; 4 \; \; 5$$
$$ | \; \; | \; \; | \; \; \; (\; \; \;)$$
$$ 1 \; 2 \; 3 \; \; 4,5$$
$$[\; k \; æ \; t \;] \; [[\; \; s \; \;]]$$

Some other candidates, ranging from promising to unlikely to hopeless, include the following.

(4.49)   Some other candidates for *cats'*
UR: $/ k_1 \; æ_2 \; t_3 / \; / z_4 / \; / z_5 /$

a. $[\, k_1 \, æ_2 \, t_3 \,] \, [\, s_4 \,] \, [\, ə \, z_5 \,]$ e. $[\, k_1 \, æ_2 \, t_3 \,] \, [\, s_4 \,] \, [\, s_5 \,]$

b. $[\, k_1 \, æ_2 \, t_3 \,]$ f. $[\, k_1 \, æ_2 \, t_3 \,] \, [\, s_5 \,]$

c. $[\, d_1 \, a_2 \, g_3 \,] \, [[\, z_{4,5} \,]]$ g. $[\, k_1 \, æ_2 \, t_3 \,] \, [\, s_4 \,]$

d. $[\, k_1 \, r \, æ_2 \, b \, b \, ɪ \, t_3 \,] \, [\, s_5 \,]$ h. $[\, s_5 \,] \, [\, t_3 \, æ_2 \, k_1 \,] \, [\, s_4 \,]$

The constraints we need in order to choose (4.48) over any of its competitors in (4.49) include some of those we saw at work with the regular plural, e.g., *SIB–SIB, VOICING, PL–AFTER–N, LEFT–ANCHOR$_{plural}$. We also require a constraint like LEFT–ANCHOR$_{plural}$ which disprefers [əz] as the realization of the possessive:[6]

(4.50)   LEFT–ANCHOR$_{poss}$: The leftmost segment of the possessive morpheme corresponds to the leftmost segment of its UR.

---

[6]There is probably a more general LEFT–ANCHOR constraint that applies to all suffixes, or perhaps all morphemes, in English.

The possessive suffix may, but need not, occur immediately after the noun stem. It's more accurate to say that it occurs at the right edge of its noun phrase. An alignment constraint to express this is:

(4.51)   POSS–IN–NP: ALIGN (Poss, Right; Noun Phrase, Right)

We now put it all together in a tableau for *cats'*.

(4.52)   **Tableau for *cats'***

/cat, pl, poss/	*SIB–SIB	POSS–IN–NP	LEFT–ANCHOR$_{poss}$
$[kæt]_N [s_4]_{PL} [s_5]_{POSS}$	*!		
$[kæt]_N [s_4]_{PL} [əz_5]_{POSS}$			*!
☞ $[kæt]_N [[s_{4,5}]_{PL}]_{POSS}$			

We still need some kind of concatenation constraint for the possessive suffix. While possessive *s* will merge with the *s* of the plural suffix, it won't merge with the final *s* of the noun stem in *mouse's*, even though this candidate, illustrated in (4.53), would be even more optimal according to the constraints we've seen so far. In the tableau in (4.54), the "☉" indicates the candidate which is incorrectly chosen as the winner:

(4.53)   / m a w s / / z /$_{POSS}$
            1 2 3 4   5
            | | | (   )
            1 2 3   4,5
            [m a w [s]]

(4.54)   **An inadequate tableau for *mouse's***

/mouse, poss/	*SIB–SIB	POSS–IN–NP	LEFT–ANCHOR$_{poss}$
$[maws_3]_N [s_4]_{POSS}$	*!		
$[maws_3]_N [əz_4]_{POSS}$			*!
☉ $[maw [s_{3,4}]_{POSS}]_N$			

To solve this problem, we might be tempted to dive straight in and create an alignment constraint analogous to PL–AFTER–N:

(4.55)   POSS–AFTER–N (to be revised): ALIGN (Poss, Left; Noun stem, Right)

Before we do, we should note that a similar contrast in merging behavior occurs after relative clauses. Stemberger (1981) observes that the possessive suffix merges with another –*s* suffix at the end of a relative clause, for example, with the third person singular verb agreement marker in:

(4.56)   The guy who cheats' [čits] friends refuse to play with him anymore.

(4.57)   / č i t /$_V$ / z /$_{3SG}$ / z /$_{POSS}$
           1 2 3   4   5
           | | | \   )
           1 2 3   4,5

           [č i t]$_V$ [s   ]$_{PL, POSS}$

However, it doesn't merge with the final *s* of a stem:

(4.58)   The guy with the ro<u>se's</u> car was towed
          [roz][əz]   *[ro[z]]

If the operative constraint were (4.55), the optimal form in (4.58) would be [ro[z]], since that allows the possessive to be one segment closer to the distant noun stem *guy* than in [roz][əz].

The constraint in (4.55) is not far from being correct. The possessive suffix does prefer to be after some stem. It's just undiscriminating about which stem it is. The noun stem of the possessor might be more obvious, but the verb stem *cheat* in (4.56) will serve just fine. So will the stem of some other noun, like *rose* in (4.58). The revised alignment constraint is:

(4.59)   POSS–AFTER–STEM: ALIGN (Poss, Left; Stem, Right)

The tableau for *(the guy who) cheats'* would be:[7]

(4.60)   **Tableau for *cheats'***

/cheat, pl/	*SIB–SIB	POSS–AFTER–STEM	LEFT–ANCHOR$_{poss}$
[čit]$_V$ [s$_4$]$_{3SG}$ [s$_5$]$_{POSS}$	*!	*	
[čit]$_V$ [s$_4$]$_{3SG}$ [əz$_5$]$_{POSS}$		*!	*
☞ [čit]$_V$ [[s$_{4,5}$]$_{3SG}$]$_{POSS}$			

This discussion of English illustrates how powerful OT can be. Only a handful of straightforward correspondence and Generalized Alignment constraints, whose effects can be seen in the simple environments, builds an account of a rather complex system of behavior in a wide range of environments.

---

[7]One consequence of this analysis that we might not want is that it forces us to treat the preposition *to* in *the guy I was talking to's car* as a Stem. Another possibility is that the base of the possessive's affixation, and the second category of the Alignment constraint in (4.59), is not a Stem but a Prosodic Word. While these kinds of sentences begin to stretch the limits of grammaticality judgments, some evidence in favor of the alternative, at least in my own speech, is that a *to* which has a *'s* cliticized to it cannot be reduced:

    The guy I was talking to [tu:/tə] won't come.
    The guy I was talking to's [tu:z/*təz] car was just towed.

Also, I find it extremely strange to merge an *'s* with a 3sg suffix which is an integral part of its Prosodic Word, although the unmerged version is worse:

    If this guy doesn't pay his bills, I'll work on the guy who does' [??dʌz/*dʌzəz] car.

Subregular Past Tenses

English may also have some less regular manifestations of the idea that more than one morpheme can share the same space in a representation. As noted in (4.5), most irregular past tenses of English resemble regular past tenses at least to the extent that they end in an alveolar stop:

(4.61)   **Irregular pasts ending in alveolars:**
   t: ate, beat, bent, bit, brought, built, bought, caught, cut, felt, fought, got, ...
   d: bound, bled, bred, could, did, fed, found, had, heard, ...

These forms do not have the alignment properties of the regular past tense forms. They cannot be cleanly divided up into pieces. If we insisted on trying to divide *fed*, we'd end up with [fɛ] and [d]—not a desirable state of affairs from the point of view of the 'feed' morpheme, since [fɛ] is quite distinct from [fi:d], but the past tense morpheme has nothing to object to: [d] is a perfectly acceptable version of the past suffix. Irregular pasts that end with what looks like a legitimate past tense morpheme outnumber those that do not by about a two to one margin. This is unlikely to be an accident. In OT, we can have the situation reflected in the correspondence relations between the irregular past form and the URs:

(4.62)   / f i d /ᵥ / d /_PAST_
            1 2 3   4
            | | (   )
            1 2   3,4
          [f ɛ   [d]_PAST_]ᵥ

It is not immediately clear what the best way is of making the irregular pasts of English look like (4.62) whenever possible, while keeping the regular pasts from ever following suit. But at least the ideas of Correspondence Theory give us a beginning point for approaching a situation which is fundamentally incompatible with the classical picture.

In this section we have seen how OT can be used in analyzing some fairly complicated aspects of English morphology, as well as some of the insights OT offers which are not easily available in a classical approach. In looking at the subregularities in past tense formation, we have also been able to glimpse some of the challenges that will have to be faced in order to extend smallish analyses like these and reach the goal of providing a complete OT grammar for an entire language.

5 Other Questions and Directions

In this final section, I look at some of the issues still remaining in the study of morphology within Optimality Theory. Some of the issues, such as the overall organization of a grammar and the place of morphology in it, have always troubled linguists. OT offers no magic answer for them, and there is no reason to assume that only one stand on the unresolved questions will be compatible with the ideas of the OT research program. Other issues, however, seem to be tailor–made for the ideas of OT, although the infrastructure for implementing an analysis may not yet exist. Most of the following remarks are speculative and only suggest some of the directions that the study of OT and morphology may move in.

The Place of Morphology

There has perhaps never been consensus on the place of morphology in grammar. Within the generative tradition, there has been disagreement over whether there is even a need for a morphological component in the same sense that there is a phonological or a syntactic component, or whether all the work of assembling words can be performed by independently needed principles of semantics, syntax, and phonology. (There is not even agreement on whether there is such an entity as a morphological word that is interestingly different from the syntactic zero–bar level or the phonological level known as the prosodic word.) Among those researchers who do believe in a morphological component, there is little agreement on how it interacts with the other components.

OT is no exception to this situation. It is often tacitly assumed that there is a morphology–like component which chooses the right underlying representations and ships them off to GEN in the phonological component, complete with handy morphological annotations like "Prefix" or "Stem", but little effort has been spent on figuring out what this component is or how it works. In fact, while there is a growing body of work in OT syntax and OT phonology, there are still few clear ideas about how they relate to each other. Is there a classical serial relationship between the two, with an OT syntax first calculating the optimal syntactic representation, which then serves as the input to an OT phonology (perhaps stopping off at an OT morphology component in the middle)? Or is there some larger, integrated grammar, where EVAL chooses all at once the best overall combination of a phonological, a syntactic and a semantic representation? Various stands on these issues are possible, and are beginning to be developed within OT, each of which will involve a different conception of the role of morphology.

An example of an unresolved question which will depend on one's overall approach to morphology and grammar is the difference between inflection and derivation. Inflectional morphology marks a noun, verb, or adjective for categories like person, number, case, gender, and tense, properties which may be syntactically relevant in fitting the word into a sentence. (The English and Paamese examples we looked at were all examples of inflectional morphology.) Derivational morphology changes a stem into a different stem, often changing its part of speech in the process and often adding some unpredictable, non–compositional element to the meaning, e.g., *nation* → *national* → *nationalize* → *nationalization*. There is some evidence that inflectional and derivational morphology are fundamentally different, each operating according to different principles, but any attempt to separate them completely runs into problems with the many cases that inhabit the middle ground, showing some properties of each.

Some of the most convincing arguments for morphological categories and principles that are independent of both syntax and phonology fit in well with the fundamental ideas of OT. For example, Aronoff's (1993) proposals for some principles of an independent morphology rely heavily on default rules. The Latin noun *nauta* 'ship' is syntactically masculine—for example, any adjective that agrees with it has to be masculine—yet the suffixes that *nauta* itself takes are those of the most common class of feminine nouns. Aronoff argues that the behavior of nouns like *nauta* cannot be accounted for using only syntactic properties (such as masculine gender) or phonological properties (such as ending in *–a*). Instead, there needs to be a purely morphological property, that of belonging to the inflection class that Latin textbooks call the "first declension". There are, however, important default mapping relations between the different kinds of properties: nouns that belong syntactically to the feminine gender typically belong morphologically to the first–declension inflection class, and vice versa. The kinds of mapping rules that Aronoff proposes to capture both special cases like *nauta* and the default case can be easily rephrased in terms of OT hierarchies where morpheme–specific constraints (if *nauta*, then first declension) outrank and can force violations of general constraints (if first declension, then feminine). It remains to be seen if individual analyses like these can be incorporated into a full–fledged OT–based morphological component, and how such a component would relate to the other components.

> Default mappings between syntactic and morphological categories can be handled naturally by EVAL.

More on Morpheme Ordering

We've seen that simple generalizations about morpheme order can be handled in a simple and natural way using constraints built with the Generalized Alignment schema. We occasionally run across some more complicated ordering systems, where the position of an affix depends on other factors in its environment, as with the prefix/infix alternation in Tagalog. An extreme example is Huave, a Hokan language of Southern Mexico (Stairs & Erickson 1969, Matthews 1972b), where certain morphemes can be prefixes in some inflected forms and suffixes in others. It would be interesting to see whether these cases can be analyzed as minimal violations of Generalized Alignment constraints in the phonology or whether a separate level of morphological representation is necessary to account for such data.

Sometimes an affix appears to be displaced from the position where we might expect to find it, not because it is there in other forms of the word, but on more general theoretical grounds. Bybee (1985) has proposed a universal ordering of the affixes that mark the verbal categories of tense, aspect, and mood: aspect markers will occur closer to the verb stem than tense markers, which in turn will be closer than mood markers. Bybee argues that a category's universal distance from the verb stem reflects how much it affects the meaning conveyed by the verb. There are, of course, exceptions to this claim of universal order. Perhaps Bybee's generalizations can be expressed as a set of Generalized Alignment constraints that form part of Universal Grammar, which may sometimes be violated when forced by some higher ranked constraint.

> Universal tendencies in affix ordering may be explained by universal alignment constraints.

The Derivational Residue

While OT has had considerable success in giving alternative analyses for those phonological phenomena which once were viewed as requiring step–by–step derivations, it remains to be seen if all vestiges of derivationalism can be done away with. One piece of derivational machinery, the idea of **cyclicity**, has been the focus of much recent work. There were many phenomena which Generative Phonology dealt with by cyclic rule application; that is, by having the same set of phonological rules apply to successively larger stretches of the same representation, the stretch affected on each cycle including at least one more morpheme than the previous cycle. Many of these phenomena have been successfully reanalyzed using OT machinery like Generalized Alignment. Other phenomena may require correspondence relations between candidates and some other independent word of the language.

Benua (1995) offers an OT analysis of an apparently cyclic effect: the tensing of found in some New York and Philadelphia dialects of English. The tensed version, represented by [E], is usually associated with closed syllables, as in the first column, and when a suffix causes the syllable to become open, the vowel is a normal [æ]. But with another class of suffix, the vowel is tense even though the syllable is open:

(4.63)	Unaffixed	Class 1 Affix	Class 2 Affix
	class [klEs]	classic [klæ.sɪk]	classy [klE.si]
	mass [mEs]	massive [mæ.sɪv]	massable [mE.sə.bl̩]

This could be handled by a cyclic analysis: the tensing rule applies after class 1 affixes have been added, but before class 2 affixes have been added (or become visible to rules). At this point in the derivation, the words in the Class 1 column have open syllables, but the words in the Class 2 column still look like those in the unaffixed column, have closed syllables, and undergo the tensing rule.

Benua proposes an OT analysis based on the idea that there can be correspondence relations between two independent surface forms. In this case, the base that class 2 affixes are added to (e.g., the [klEs] of *classy*) has its correspondence relations with the unaffixed form in the first column (e.g., [klEs]) rather than directly with the UR (e.g., /klæs/). Faithfulness will require its low vowels to be tense, even though its open syllable should be unaffected by the tensing constraint. This is just like the way faithfulness between reduplicant and base forced the Paamese base to be *luhu*, even though it would otherwise have been unaffected by the *$i+u$ constraint. The empirical predictions of allowing correspondence relations like this between surface forms remain to be explored.

> Correspondence relations between two surface forms provide a way of explaining "cyclic effects".

An interesting twist on having faithfulness between surface forms is the reverse situation: where two surface forms are systematically *different* from each other. Situations involving systematic *un*faithfulness crop up occasionally in the OT literature. Fitzgerald & Fountain (1995) propose a constraint for Tohono O'odham that the perfective form of a verb be shorter than its imperfective form (causing, for example, imperfective *nákog* 'enduring' to shorten to *náko* in the perfective). A similar constraint seems to be tacitly assumed in Benua's discussion of Icelandic truncation. So far there is no sign of a general theory of the nature and limits of unfaithfulness constraints.

One Form, One Function

Languages seem to strive toward, but always fail to reach, a state of perfect one–to–one correspondence between sound patterns and meanings, or forms and functions. Languages will sometimes go to great lengths to avoid situations of homonymy (a single phonological representation with more than one meaning) or synonymy (two or more phonological representations with the same meaning), although there are not yet any ways of predicting when a language will rouse itself out its usual apathy in this respect.

Truncation can be seen as a special case of homonymy avoidance: Tohono O'odham perfective and imperfective are different functions, they should optimally have different forms. Synonymy avoidance in the domain of morphology is usually referred to as "blocking" (cf. Aronoff 1976). For example, the English suffix *–ity* can derive an abstract noun from an adjective, as in (4.64a), but the derivation will be systematically blocked if there is already an abstract noun with the same meaning (4.64b–c).

(4.64) a. curiosity, luminosity, ...
        b. *gloriosity, *furiosity, *acrimoniosity
        c. glory, fury, acrimony

Cases where synonymy seems *not* to be avoided also pose interesting challenges for OT (e.g., Anttila 1995). Equally acceptable forms, like the English past tense forms *dived* and *dove* for many speakers, may or may not prove to be solvable by devices like allowing constraints to remain unranked with respect to each other.

Besides the sporadic, but frequent, influence that the ideal of a one–to–one form–function mapping can exert on the grammaticality of linguistic items, it appears to be an important strategy in language acquisition to assume that the ideal actually holds (Clark 1987). It may be that the ideal that languages are fumbling for will be an important constraint or family of constraints in an OT–based theory of the mapping between phonology, syntax, and semantics—often dominated and overruled, but always present, waiting for the chance to exert itself.

> Is the subtle tendency for one form to have one function (and vice versa) the result of low–ranking universal constraints?

Paradigms

Textbooks of languages like French will typically have at the back page after page of paradigm tables showing the various forms of selected verbs. Linguists have often treated the paradigm, the structured collection of the possible forms of a stem, as a descriptively convenient fiction. But there is growing evidence that paradigms may be independent (perhaps primitive) entities that morphological theory will have to come to terms with (see, for example, many of the papers in Plank 1991). Carstairs (1987) proposes a Paradigm Economy Principle to account for the puzzling scarcity of the creatures: if German has two possible accusative singular suffixes, four for the genitive singular, three for the dative singular, and five each for the nominative, accusative, and genitive plural, and four for the dative plural, then it would be logically possible for German to have $2\times4\times3\times5\times5\times5\times4 = 12,000$ possible noun paradigms. Instead of the theoretical maximum, German noun paradigms number under a dozen, and, with enough clever reanalysis, achieve the theoretical *minimum* of five. Carstairs argues that paradigms are real and that universal constraints minimize how many of them a language can have. Constraints on the structure of paradigms, many of them versions of the one form–one function ideal discussed earlier, have also been proposed as acquisition strategies for inflectional morphology (e.g., Pinker 1984). OT's ability to set up correspondence relations between two independent surface forms (e.g., between any inflected form of a verb and the third person singular form) may go a long way towards explaining many of the interesting properties of paradigms, although the infrastructure for constructing convincing analyses does not yet exist.

> OT correspondence relations between inflectionally related words may explain many of the properties of paradigms.

Acquisition

I have touched on acquisition at various points in this discussion. One leading idea on how an OT grammar could be acquired which has proven useful in both theoretical (e.g., Tesar 1995) and empirical discussions (e.g., Gnanadesikan 1995) is that the initial state of a grammar has the universal unmarkedness constraints very highly ranked—these universal constraints can be progressively demoted in the face of counter–evidence (e.g., adult candidates that violate them), or alternatively faithfulness constraints can be promoted, until a grammar is reached which selects the same winning candidates as the rest of the speech community. If this is an accurate picture of the language acquisition process, then the kinds of strategies and assumptions that we find children using in learning the morphology of their language should not be treated as secondary facts of performance with no necessary relation to linguistic competence.

> The strategies and assumptions children use in acquiring language may be direct manifestations of universal constraints which continue to be present in the hierarchy of the adult's grammar.

5

Optimality Theory and Syntax: Movement and Pronunciation*

David Pesetsky

The notion of "optimization" standardly assumed by Optimality Theory is quite simple and natural. While not uncontroversial, it has advanced our understanding of many long–standing problems in phonology. Nonetheless, it is not the *only* simple and natural proposal to enter linguistic theory. The successes of OT within phonology and mor-phology) do not *necessarily* teach us anything more general about how other aspects of language work. On the other hand, they might.

It would be interesting, for example, if we were to discover that the mechanics of constraint interaction proposed by Prince and Smolensky (1993) characterize *all* aspects of linguistic knowledge and use. Such a discovery would reveal a striking uniformity across linguistic processes that would immediately raise wider questions about possible uniformity of interaction beyond the boundaries of language. Is the human language faculty, or the mind, more broadly, an optimizer?

*A short, more technical survey of the work discussed here appears in Pesetsky (to appear), but the work is part of a much larger endeavor. The presentation in this chapter is intended to illustrate some results and issues, and does not at every point present its material with the care that might constitute proof. Syntactically informed readers will recognize many cut corners and roads not taken. I am grateful for discussion of this work to audiences at the University of Arizona, Eotvos Lorand University, the Generative Grammar Circle of Korea, the Holland Institute of Generative Linguistics, the Numazu Summer Seminar in Linguistics, Sophia University, DIPSCO at Ospitale San Raffaele, the University of Paris-Nanterre, the University of Pennsylvania, UCLA and USC, as well as to students at MIT. Particularly helpful were discussions with Noam Chomsky, Danny Fox, Paul Hagstrom, Richard Kayne, Martha McGinnis, Eric Reuland, Donca Steriade, Hubert Truckenbrodt and Ken Wexler. Tom Bever provided useful comments on the lecture from which this chapter was developed.

It would be equally interesting, however, if we were to learn something entirely different: that optimality–theoretic interactions characterize *some, but not all*, aspects of language. We would then face the intriguing task of discovering exactly what Optimality Theory is and is not good for — and, ultimately, understanding the reasons for any divisions of labor that we may find.

Clearly, these are issues that can only be addressed through serious investigation. These are not questions that can be given useful *a priori* answers. The initial case for OT in phonology rested on conceptual flaws in standard accounts of phenomena like Berber syllabification and the cross–linguistic character of infixation. These phenomena display such an obviously optimality–theoretic character that alternative, more traditional treatments seemed to miss the point. In parallel fashion, we should ask whether any problems of syntax have a similar character. Are there problems connected with the grammar of sentences for which traditional, non–OT treatments patently miss the point somehow. If the answer is "yes", we must also ask whether there are areas in which it is the *OT* treatment that "misses the point". Once we have a body of work that asks and answers both types of questions, we will be on our way towards a general understanding of the scope and nature of optimization phenomena in human language.

This chapter and the next sketch what such an investigation might look like in the domain of **syntax**. In order to pursue this goal, however, we first need a general understanding of what is known about syntax. I therefore begin with a sketch of the "state of the art", enumerating some discoveries about sentence structure that seem the most profound and important. Only once we gain some clarity of vision about the field will we be in a position to ask where (if at all) ideas from Optimality Theory are relevant.

I will suggest that there is only weak evidence for OT constraint interaction in several areas traditionally viewed as central to syntax, and in fact some serious arguments against an OT view in these domains. On the other hand, other areas of syntax — in particular those that lie at the boundary between syntax and phonology — have a strikingly obvious OT character. This is no surprise, if OT interactions are a hallmark of aspects of language closest to the mouth and ear, but is perhaps a surprise if OT is a "theory of everything". The technical discussion in this chapter will therefore bear directly on the fundamental questions with which the chapter began.

We need to start with a few words of warning. The discussion in this chapter is not only pedagogical (cutting a number of corners), but also speculative and tentative. The investigation of OT properties of syntax is in its infancy. Evidence for OT interactions might be strongest at the boundaries with phonology merely because we know more about OT in the domain of phonology — or it may be the case that OT really *is* a theory of phonology. This chapter will not decide these matters, but will try to point out how they might eventually be decided.

1 The Components of Syntax: Constituency and Movement

The history of syntactic investigation is marked by a small number of central discoveries which created the syntactician's research agenda. One can divide these discoveries into two groups — the discovery of **hierarchical constituent structure**, and the discovery

that elements may occupy more than one position within this hierarchy, which the literature calls **movement**, for reasons we shall shortly discuss.

From Grammatical Relations to Constituent Structure

One of the most important syntactic achievements, traceable to the work of pre–modern
grammarians, is the discovery of **grammatical relations**. This discovery involves, first,
the realization that words group together to form phrases, and second, that not all phrases
are created equal. Thus for example, there is a difference between phrases that function
as subject–of–the–sentence and phrases that function as object–of–the–sentence. The
discovery was understandably abetted by the encoding of such distinctions in the
morphology of languages like Greek, Latin, Sanskrit, Japanese — but crucially involved
the recognition that an *abstract* notion lay behind superficial facts of morphology.

(5.1)   **Hierarchical constituent structure**

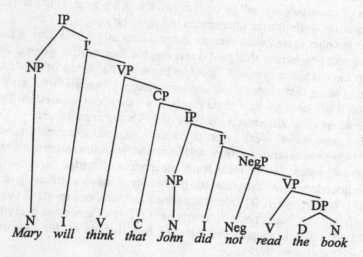

Legend:

C	=	complementizer (subordinating conjunction)
D	=	determiner (article, pronoun, possessive *'s*)[1]
I	=	inflection
N	=	noun
Neg	=	negation
V	=	verb

XP indicates maximal phrase headed by X (VP = verb phrase, etc.)
X' indicates intermediate phrase headed by X (read as "X–bar")

---

[1] In the structures drawn in this chapter, I do not assume that D and DP are present when there is no
determiner word present. That is why *Mary* is an NP, but *the book* is labeled DP. This assumption
might not in the end be correct, but will suffice for our purposes.

In this way, for example, languages that lack rich morphology could be insightfully described using the same categories that make obvious sense in languages with rich morphology, and a general understanding of human–language syntax could be developed.

A related discovery was the realization that the phrasal organization of sentences is **hierarchical** — generally characterizable by the sort of **phrase–structure** tree exemplified in (5.1) for the English sentence *Mary will think that John did not read the book*. Thus, while words *the book* are a noun phrase (NP) in (5.1), they also form part of a larger phrase that properly contains it — the verb phrase (VP) *read the book*. That verb phrase, in turn, forms part of larger phrases *not read the book, did not read the book*, etc. — each of which in turn may be composed of a set of phrases. One element of each phrase, called its **head**, determines its major properties. Phrases are **labeled** to reflect the identity of their heads; see the Legend to example (5.1) above.

Constituency reveals itself in many ways. For example, the constituents of a sentence are the fragments that can be uttered in surprise.[2] So, if I never knew the fact communicated by sentence (5.1) before, I might exclaim in surprise (5.2), repeating the constituent labeled **I'**. But I can't utter in surprise (5.3), picking out a nonconstituent.[3]

(5.2)   Did not read the book?? You must be kidding!

(5.3)   *Did not read the?? You must be kidding!

Syntactic constituents bear a law–like relation to semantic chunks (though the correspondence is often more complex than one–to–one). For example, the fact that the proposition *that John did not read the book* functions as an argument of the verb *think* corresponds to the fact that this sequence forms a syntactic constituent.

In addition, the verb of the sentence combines with the direct object to form a constituent that excludes the subject. Consequently, verb+object, but not verb+subject, can be a sentence fragment uttered in surprise.

(5.4)   Sue unwrapped the present.
    a. Unwrapped the present?? You must be kidding!
    b. *Sue unwrapped?? You must be kidding!

This fact about verbs and direct objects — a possible universal fact about languages — will be important in the next section.

Constituent structure also places limits on the ways in which phrases may **corefer**. In (5.5a), we cannot understand the pronoun as referring to the same individual as *John*. (That is, *John* cannot function as the **antecedent** for the pronoun.) On the other hand,

---

[2]I owe the "sentence fragment test" for constituency to Radford (1988), a standard textbook covering some of the fundamentals discussed in this chapter. A more comprehensive survey is provided by other textbooks, especially Haegeman (1994).

[3]The fragment *I don't remember the* naturally occurs when the speaker can't think of the word that follows *the*. This is not a counterexample because the speaker intended to utter a constituent.

(5.5b) is different. Here the pronoun *his* may refer to *John*. (*John* may function as the antecedent for *his*.)

(5.5)   **Patterns of coreference and non–coreference**
    a. He will criticize the picture of John.
        [*He* cannot be *John*]
    b. His former teacher will criticize the picture of John.
        [*His* may be *John*]

The source of this difference does not lie in the fact that the pronoun precedes its would–be antecedent, since this is true of both examples. Instead, the source of the difference is structural, namely that the pronoun *he* in (5.5a) is attached to the tree at a node that contains its antecedent *John*. That is, the first available **sister** node to *he* is the phrase *will criticize the picture of John*, and that phrase contains *John*. Syntacticians call this relation **c–command**. We say that *he* c–commands *John* in (5.5a).

(5.6)   **Structure of (5.5a)**

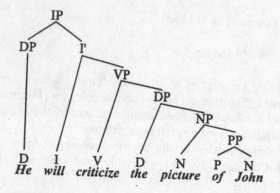

Legend:
P   =   preposition

By contrast, the first sister node available to the determiner *his* in (5.5b) is the NP *former teacher*, which does not contain *John*. Consequently, *his* in (5.5b) does not c–command *John*. This means that *John* can function as the antecedent for *his*.[4] Examining many contrasts of this sort, linguists have generally concluded that coreference between a pronoun and a non–pronoun is restricted by c–command relations of just this sort. A pronoun may precede its antecedent, but: may not c–command it.

---

[4]Note also that ours is a fact about *structure*, not about the difference between *he* vs. *his*. For example, in *his picture of John*, the pronoun *his* cannot corefer with *John* because it c–commands it, even though *his* and *John* may corefer in (5.5b).

(5.7)   Structure of (5.5b)

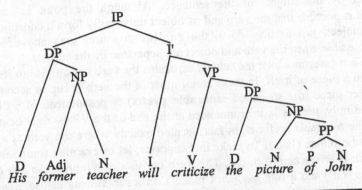

> The fact that so many phenomena refer to constituency — the grammar of sentence fragments, semantic interpretation, limitations on coreference, and much more besides — serves as evidence that constituent structure is a central property of syntax.

Movement

The second central property of syntax that we examine is the discovery that some phrases and words bear *more than one* grammatical relation, i.e. that *some phrases and words occupy more than one structural position* in the hierarchical structures just discussed.

Consider, for example the rules that govern the placement of the verb in standard German. Every foreign student of German has run afoul of the fact that the position of the German verb depends on the type of clause it occupies. A common description of the situation is the pair of statements in (5.9).

(5.9)   German verb placement: a common description

    a. In a main clause, the tensed verb follows the first phrase.
      ["verb–second"]

    b. In a subordinate clause, the tensed verb comes last.
      ["verb–final"]

These statements accurately describe the differences the placement of **sah** in (5.10a–b).

(5.10)   a.   Gestern   **sah**   Hans den Mann.
            yesterday saw Hans the man
            'Yesterday, Hans saw the man.'
            [main clause, verb–second]

      b.   Marie denkt,   dass Hans gestern   den Mann **sah**.
            Mary   thinks,   that Hans yesterday the man   saw.
            'Mary thinks that yesterday Hans saw the man.'
            [subordinate clause, verb–final]

From a cross–linguistic perspective, (5.10a) is a puzzle, since the verb is separated from its object by the subject of the sentence. Although the point is somewhat controversial, it is possible that the verb and its object universally form a constituent that excludes the subject, just as they do in the English structures seen above.[5] This is obviously not possible when the verb and object are separated by the subject.

In fact, there are cases in which the subject separates the verb not only from its object — but also from a piece of itself. In these cases, most of the verb occupies second position, and another piece (the so–called **separable prefix**) is pronounced in a different position. Interestingly, the prefix is pronounced at the end of the VP — right next to the direct object. The separable prefix is, in fact, located exactly where the verb as a whole would sit in a subordinate clause. To make this concrete, let us examine some sentences whose main verb is *anmachen* 'turn on'. This verb consists of a morpheme *an–* (which otherwise means 'on') and a root *machen* ('make'). In a subordinate clause, *anmachen* shows up in one piece in V–final position. In a main clause, however, only the morpheme *machen* shows up in second position. The morpheme *an* shows up in final position, i.e. the spot where the whole verb is pronounced in subordinate clauses:[6]

(5.11) a.   Wir –machen jetzt das Licht   an–.
              we   make   now   the light   on
              'We are now turning on the light.'
              [main clause, verb–second]

        b.   Marie denkt, dass wir jetzt das Licht anmachen.
              Mary   thinks that we   now   the light on–make
              'Mary thinks that we are now turning on the light.'
              [embedded clause, verb–final]

It is as if the verb *anmachen* had its pronunciation spread out over two separate structural positions — the "normal" verb position at the end of the sentence next to the direct object, and the verb–second position near the front of the sentence. In this way, we see a situation in which a single word *anmachen* seems to occupy two positions at once — with pieces of the word pronounced in each position.

In fact, we can say more about these two positions. The verb–final position of subordinate clauses is the expected position for the verb in a language where the heads of phrases are final in their phrase. There are many languages in which the verb is final within the verb phrase — for example, Japanese (see Chapter 6).

(5.12) Hanako–wa   susi–o   tabeta.
        Hanako–TOPIC sushi–OBJ ate
        'Hanako ate sushi'

---

[5]Some recent work develops this idea in a manner different from that seen in (5.1). I have presented the older, structurally simpler theory for expository reasons.
[6]The significance of these cases was first noticed in the early 1960s by Bach (1962) and by Bierwisch (1963).

The nature of the verb–second position is less obvious and more interesting. Den Besten (1983) was perhaps the first to point out the significance of some important classes of exceptions to the standard description of German word order in (5.9) — certain subordinate clauses with verb–second order. For example, in addition to (5.13a) with the expected verb–final order in the subordinate clause, (5.13b) is also possible. Example (5.13b) shows verb–second word order in the embedded clause, as well as an unexplained change of mood to subjunctive. Likewise, in addition to (5.14a), we also find (5.14b), which once again shows verb–second word order in the embedded clause. The clauses of interest in the following examples are therefore the subordinate clauses, in which the verb of interest is highlighted in boldface.

(5.13)   **Main clause word order in embedded clauses**
    a. Hans sagte, dass er glücklich **ist**.
    Hans said   that he happy   is
    'Hans said (that) he is happy.'
    [verb–final]
    b. Hans sagte, er **sei** glücklich.
    Hans said   he be happy
    [verb–second]

(5.14)   **a.** Er benahm sich,   als ob er noch nichts   gegessen **habe**.
    he behaved himself as if he yet   nothing eaten   had
    'He behaved as though he had eaten nothing.'
    [verb–final]
    b. Er benahm sich,   als **habe** er noch nichts   gegessen.
    he behaved himself as had   he yet   nothing eaten
    [verb–second]

What induces verb–second is not main clauses, but clauses that are not introduced by a complementizer. That is why verb–second order shows up when *dass* is missing in (5.13b) and *ob* is missing in (5.14b). Consequently, it is tempting to suggest that the verb in verb–second clauses in fact occupies the complementizer position. The same conclusion can be drawn from (5.13b), with one caveat. The verb is in second, rather than first position within CP, because of the existence of one extra position to the left of the complementizer. This position is often called the "specifier of CP" — or **SPEC(CP)**. The topic of the sentence occupies SPEC(CP), as do certain other elements, including clause–initial adverbs. Many or all phrases seem to have a unique specifier position of this sort. For example, the subject of the sentence is the SPEC(IP). Some people also suggest that the possessor phrase in a nominal (*Bill*'s *book*) is the SPEC(DP) — with the *'s* occupying the D position. Den Besten's hypothesis thus holds that the verb in German *occupies two positions* in certain clauses, and that the position to the left of the verb in verb–second clauses is the SPEC(CP).
In other words, German only *seems* to display a difference between main vs. subordinate clauses. The reason is the interaction of the "better description" in (5.15) with the entirely independent fact that main clauses are not usually introduced by a subordinating conjunction. The grammatical representations for (5.10a–b), then, are (5.16a–b), with the

shaded box indicating the complementizer position — containing the complementizer in (5.16a), and the tensed verb in (5.16b). The line through the second verb–final occurrence of *sah* in (5.16a) indicates that the verb occupies this position, but is not pronounced there.

(5.15)   German verb placement: a better description
   a. The verb occupies the rightmost position within the VP.
   b. When the complementizer of a clause is not pronounced, the finite verb of the clause not only occupies the rightmost position within the VP, but also occupies the position of the complementizer.

(5.16)   German verb in the complementizer position

   a. Gestern | sah | Hans den   Mann ~~sah.~~
       Yesterday | saw | Hans the   man   saw

   b. Marie denkt, | dass | Hans gestern   den   Mann sah.
       Mary thinks, | that | Hans yesterday the   man saw

One important note about this phenomenon. For our purposes, all we need to assume is the concept just developed: the idea that certain words and phrases are associated with more than one structural position. Nonetheless, there is some evidence supporting a more elaborate view, according to which the positions that a phrase occupies are arranged in a sequence. According to this view, assignment of a phrase to more than one position is the consequence of a **movement** procedure that copies a word or phrase into the various positions of this sequence.[7] The view of verb–second in German as *movement of the finite verb to the complementizer position* is exemplified in (5.17).[8]

---

[7]Chomsky's (1995b) **Minimalist Program** (see Chapter 6) adopts the sequential view not only for movement, but also for the laws that govern basic constituent structure. The fundamental objects of syntax in this theory are constructed by the rules **Merge** (form constituent) and **Move**. Nothing in this chapter bears on the correctness of this view, often called "derivational". OT work in phonology generally does not motivate derivational notions as part of constraint evaluation, but does leave open the possibility of a derivational character for GEN, the procedures that yield the candidates that the constraints evaluate.

[8]Some notes on (5.17). First, the adverbial in SPEC(CP) also occupies that position by virtue of movement. Second, the I node here is not filled by an auxiliary verb (like have), but is probably also a position occupied by the main verb. Movement of V to C is actually movement of V to I, and then to C. Finally, in other work (Pesetsky to appear), I argue that the verb in C is actually attached to the left of an unpronounced complementizer, rather than occupying the actual complementizer position as in (5.17). If you are reading this chapter for the second time, you can see for yourself that the discussion developed below explains why the complementizer must be unpronounced. These details can be ignored for present purposes.

(5.17)   **Movement of V to C**

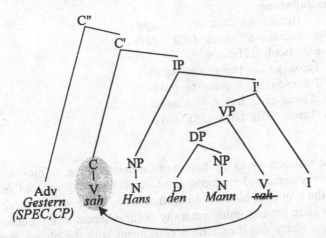

Since the term "movement" is so common, I will adopt it here — but the sequential aspect of this description will play no role in our discussion.[9] What is important to us about movement is the basic discovery that a word or phrase may be associated with more than one position.

Movement phenomena abound in the languages of the world. Furthermore, every now and then, we are lucky enough to find instances of movement in which, as with German prefixed verbs, one can actually hear the pronunciation of a syntactic unit "spread across" more than one syntactic position. Japanese furnishes another telling example. The Japanese verb and its inflections are quite rigidly located at the end of the sentence,[10] but in many other respects, word order is quite free. For instance, the Japanese rendering of *Taroo gave the book to Hanako* may show any order of constituents — so long as the verb remains at the end. This sort of freedom is called **scrambling**.

Armed with our understanding of constituent structure and movement, we might be tempted to attribute this variation in word order to movement. Such a proposal would allow us to analyze Japanese as a language with a rich constituent structure and a simple fundamental word order: Subject – Indirect Object – Direct Object – Verb. Movement of the subject and objects would explain the fact that subject and object sometimes show up in unexpected positions. But then remorse and guilt might set in. "How Anglocentric of us to impose on Japanese the rigidities of English word order," we might think.

---

[9]Some schools of thought in modern syntax assume a view of "movement" phenomena in which significantly less information may be shared among the various positions than in the model discussed in this chapter. For example, the research tradition called first **Generalized Phrase Structure Grammar** (GPSG; see Sells 1985; Gazdar, Klein, Pullum & Sag 1985) and later **Head–Driven Phrase Structure Grammar** (HPSG; Pollard & Sag 1994) differs from our proposal in this manner. The reader is invited to consider how these alternative traditions might handle the evidence presented in this chapter in favor of full multiple association (movement), but I will not undertake this sort of comparison here.

[10]There are arguments that the Japanese verb moves from VP–final position rightwards to a CP–final position (Koizumi 1995). These arguments do not affect our discussion.

(5.18)   **Scrambling in Japanese**
  a. Taroo–ga   Hanako–ni   hon–o   ageta.
     Taroo–SUB   Hanako–to   book–OBJ   gave
     'Taroo gave the book to Hanako.'
  b. Hon–o   Taroo–ga   Hanako–ni ageta.
     book–OBJ   Taroo–SUB   Hanako–to gave
  c. Hanako–ni Taroo–ga   hon–o   ageta.
     Hanako–to Taroo–SUB   book–OBJ   gave
     etc.

Nonetheless, our first instincts — to attribute word order freedom to movement — might be correct, and our remorse and guilt misplaced. Kuroda (1983) reported an interesting discovery about the syntax of numeric quantification in Japanese. Japanese numbers, when they modify noun phrases, must generally occur adjacent to the phrases that they modify. Indeed, it is likely that they form a constituent with the phrase that they modify. Just as *an–* in German *anmachen* is expected to appear next to *–machen*, so one expects the *two* in *two students* to appear next to *students*. In simple sentences, this expectation is confirmed, as we can see in (5.19). The examples are carefully constructed so that we can test this expectation against easy judgments of acceptability. The **classifier** *ri* on the numeral 'two' in (5.19) agrees with *gakusei* 'student', making it clear that it is students (not pizzas) that are being counted. It is not acceptable to separate *futa–ri* from *gakusei*.

(5.19)   **Japanese numerals adjacent to the phrases they modify**
  a. **Gakusei–ga   futa–ri**   piza–o   katta.
     students–SUB   two$_{student}$ pizza–OBJ   bought
     'Two students bought pizza.'
  b. *****Gakusei–ga** piza–o   **futa–ri**   katta.
     students–SUB   pizza–OBJ   two$_{student}$ bought

The reason for this contrast should be clear. Since the numeral and the phrase it modifies form a semantic unit, they should form a syntactic unit. When they are separated by other material as in (5.19b), they cannot form a syntactic unit, since they cannot form a constituent.

Now consider examples in which NPs are scrambled; i.e. occur in positions that do not conform to our expectations about syntactic and semantic units. Kuroda noticed something quite surprising about such cases. In sentence containing an NP that isn't where it "should" be, the numeral that modifies it has a choice as to where it is pronounced: next to where the NP *is* or next to where that NP *should be*. Thus, for instance, if one wishes to express 'A student bought two pizzas' in Japanese with *pizza* scrambled to the *left* of the subject, the expression for *two* can be found either adjacent to *pizza* or adjacent to the verb — next to where the direct object *is* or next to where it *should be*. How can we interpret these facts?

Think back on the German verb *anmachen* 'turn on'. In clauses missing a pronounced complementizer, the two pieces of this verb (a semantic unit) are nonetheless pronounced in two different places — *machen* in the complementizer position and *an* in the place where the German verb "should be" — the end of the verb phrase. This was taken as evidence for the association of the verb with two different positions, i.e. movement from V to C.

(5.20)   **Evidence for scrambling as movement**
    a. Piza–o   **futa–tu** gakusei–ga   katta.
       pizza–OBJ   two$_{pizza}$ student–NOM bought
       'The student bought two pizzas.'
          [numeral next to where *piza* is]
    b. Piza–o   gakusei–ga   **futa–tu** katta.
       pizza–OBJ student–SUB two$_{pizza}$ bought
          [numeral next to where *piza* "should be"]

Kuroda's discovery has much the same character. In Kuroda's cases, the two pieces of the semantic unit *numeral + modified phrase* are also pronounceable in two different places. One of these places (next to the scrambled element) is unsurprising. The other (next to where the scrambled element "should be") looks like yet another instance in which a single phrase (*piza–o*) is associated with two positions, with pronunciation "spread out" over the positions. Putting it in terms of movement, we say that Japanese NPs optionally move from their expected positions (e.g. objects next to the verb) to a "scrambling position" at the left of the sentence. The syntax of Japanese numerals teaches us that scrambling is not a sign that Japanese lacks the constituent structures of English, but is, instead, an instance of movement.

Consider now the word order seen in English questions that contain a *wh*–word or phrase like *what, who, which book* or *how*. Instead of appearing in their expected position (direct object, subject, adverbial, etc.), the *wh*–element appears in a sentence–initial position which syntacticians have analysed as the same SPEC(CP) position that we noticed first in German (cf. (5.16)). In the examples below, the expected position of the italicized *wh*–phrase is represented by an underscore:[11]

(5.21)   *Wh*-phrases
    a. *What* did he get ___ for Christmas?
    b. *Who* did you see ___ when you were in Tucson?
    c. *Which book* did Mary put ___ on top of the dresser?
    d. *How* did the employees treat the workers ___?

An obvious hypothesis at this point would attribute the "unexpected" positioning of the *wh*–phrases in (5.21) to an instance of movement involving both the underscored position and the clause–initial SPEC(CP). This is, in fact, a standard hypothesis about such constructions. The movement process is called *wh*–**movement**.

---

[11]Evidence that that the phrase occupies SPEC(CP) comes from the similarity to German verb–second order in matrix questions. Once again, the finite verb appears in C to the left of the clause, with *what* or *who* appearing in a unique position to the left of the finite verb. As in German verb–second constructions, the auxiliary verb in main–clause questions in English is probably occupying the C position. I will not focus on this fact here, however.

Moreover, McCloskey (1995) provides evidence for *wh*–movement that is quite parallel to the evidence we have examined for V–to–C movement in German and for scrambling as movement in Japanese. McCloskey's evidence comes from a dialect of Irish English spoken in West Ulster.

West Ulster English, like many American dialects, can modify question words like *what* with *all* (*what all*). This usage makes *what* a plural.

(5.22) a. **What all** did he get ____ for Christmas?
       b. **Who all** did you see ____ when you were in Derry?

McCloskey noticed that the *all* of *what all*, like the Japanese numerals, can also be pronounced in the position of the underscores — i.e. in the "expected" positions of the moved phrases. In other words, the *wh*–phrases at the front of the sentences in (5.22) also have a presence in direct object position, following the verb. This is the classic pattern of movement.

(5.23) **Evidence for *wh*–movement**
       a. **What** did he get __ **all** for Christmas?
       b. **Who** did you see __ **all** when you were in Derry?

It is of course necessary to establish, as we did for Japanese in (5.19b), that this pattern is really due to movement, and that speakers of West Ulster English do not just say the word *all* wherever they feel like it. This is the point of (5.24) below. These are cases in which *all* is separated from its *wh*–word but is not pronounced in the direct object positions with which the *wh*-phrases are associated:[12]

(5.24) a. *****What** did he **all** say (that) he wanted ____ to do?
       b. *****Who** did he meet ____ in Derry yesterday **all**?

The impossibility of (5.24a–b) reinforces McCloskey's interpretation of (5.23a–b), and, with it, the concept of movement developed in this chapter.

Summary

We have now taken a quick look at two central phenomena of sentence grammar. We saw first how the words of sentences are hierarchically organized into phrases. We then saw that certain words and phrases occupy more than one position in syntactic structure — a phenomenon called *movement*. We were fortunate to be able to detect movement through the existence of constructions in which different parts of a moved phrase are pronounced in different positions that this phrase occupies. In the next section, we

---

[12]West Ulster Irish does in fact allow *all* to be pronounced to the left of complementizers in sentences in which subordinate clauses themselves contain subordinate clauses, as in **What** did he *say all (that) he wanted to do?*. What this means is that a phrase associated with the highest SPEC(CP) position may be associated with other SPEC(CP) positions in a syntactic tree. (This is sometimes called **successive cyclic movement**.) West Ulster English evidence for this proposal was the main point of McCloskey's paper.

examine in more detail some of the laws that govern syntactic structure and syntactic movement — asking the central question of this volume: how do these laws interact?

The story so far:

1. The units of syntax are hierarchically organized. That is, sentences have internal **structure**.

2. Some words and phrases occupy more than one position in these structures, a phenomenon called **movement**.

3. There are laws governing the fundamental organization of phrase structure, as well as movement. We have not yet discussed how these laws interact.

---

2 The Problem of Ineffability

In particular, we want to ask whether the pattern of constraint interaction in phonology investigated by Prince and Smolensky also governs the laws of syntax. Our survey of syntax allows us to approach this question in a fairly informed and well–grounded fashion. We can ask, for example, (1) whether the principles that distinguish proper from improper constituent structure interact in an OT manner, and (2) whether the principles that guide movement interact in an OT manner.

Ineffability Suggests a Clash & Crash Model

The predominant research traditions give some negative answers to these questions. In fact, there are good reasons for these negative answers. In many domains, it looks as though the laws underlying constituency and movement might not interact in an obviously OT fashion.

For example, syntacticians have studied the conditions which require wh–movement in English questions. They have also studied conditions which sometimes make wh–movement impossible. The adverb how, for example, can undergo wh–movement — the process we first looked at in West Ulster English. In (5.25) how has moved from a subordinate clause.

(5.25)   I'm wondering *how* you think [employees should treat their subordinates __]

But movement of *how* from other types of subordinate clauses is completely impossible. For example, movement of *how* from a relative clause is not allowed. One might think of uttering a sentence like (5.26) — for example, if you want to be reminded of the regulations governing on–the–job behavior — yet the construction, with *how* understood as modifying *treat their subordinates*, is completely impossible.

Crucially, there is *no* acceptable alternative to (5.25) that involves the same words and the same interpretation. For example, a structure in which *wh*–movement simply fails to occur is also impossible.

(5.26)   *I'm wondering *how* the company will fire [any employees who treat their subordinates __].

(5.27)   *I'm wondering the company will fire [any employees who treat their subordinates *how*].

The standard interpretation of these facts attributes them to two constraints interacting in an apparently *non*–OT fashion:[13]

(5.28)   **Obligatory *wh*–movement:** An embedded question must have a *wh*–phrase in SPEC(CP).

(5.29)   **Constraint on movement:** *Wh*–movement cannot extract an adverbial from a relative clause.

When the desires of the two constraints clash, the conflict does not seem to be resolved by ranking one constraint over the other and picking an output that satisfies the more highly ranked constraint. Rather, the conflict seems to block the existence of any acceptable output — a situation called **crash** by Chomsky (1995b). We can therefore call this traditional perspective on constraint interaction **Clash & Crash**. Evidence for Clash & Crash comes from the phenomenon of **ineffability** — inputs that seem to yield no acceptable output. OT approaches, by contrast, always pick an output for any given input, since every clash is amicably resolved through constraint ranking. Now the fact of the matter is that there is indeed no way to ask the question in (5.26) and (5.27) using the means (words and structures) employed in these examples.

Ineffability appears to be a widespread phenomenon. Let us look at another example. Although we see *wh*–movement of a *wh*–phrase to SPEC(CP) accompanied by verb movement to C itself, we do not typically see *wh*–movement to SPEC(CP) accompanied by an actual complementizer pronounced in C.

(5.30)   *I wonder *who* that Mary invited __ to the party.

For the moment, let us call this the ***Wh*–C Constraint**, although we will actually end up explaining (5.30) in a very different manner.[14]

Sometimes, however, the *presence* of a pronounced complementizer is essential to the well–formedness of a structure. For example, an infinitival clause only allows a

---

[13]These are not their standard names. The constraint requiring *wh*–movement is generally assumed to follow from a general theory about obligatory vs. impossible movement. The constraint blocking adverbial extraction is a special case of a more general constraint called the **Empty Category Principle** (ECP).

[14]The *Wh*–C Constraint is generally known as the **Doubly Filled Comp Filter**, a name that reflects some earlier analyses of the phenomenon (Keyser 1975; Chomsky & Lasnik 1977).

pronounced subject when the subject is immediately preceded by an active verb or by the prepositional complementizer *for*. In (5.33b) below, *Bill* is immediately preceded by the noun *preference*, which is the source of its unacceptability.

(5.31) **Wh–C Constraint**: No complementizer is pronounced in C when a phrase is pronounced in SPEC(CP).

(5.32) a. Sue would prefer for Bill not to tell Pete about the decision.
       b. Sue would prefer Bill not to tell Pete about the decision.

(5.33) a. Sue's strong preference for Bill not to tell Pete about the decision.
       b. *Sue's strong preference Bill not to tell Pete about the decision.

This fact probably follows from a more general requirement. This requirement, discussed further in Chapter 6, requires pronounced NPs to occupy the sort of position in which a language like Latin might make morphological case marking available. Syntacticians call this requirement the **Case Filter**. It amounts to the claim that languages without much morphology on their nouns nonetheless act as though the morphology were present.

(5.34) **Case Filter**: A pronounced NP must occupy a Case position.

The subject of an infinitive is only a case position when it is structurally close to a case–assigning element such as the prepositional complementizer *for*. This is why the pronounced subject of the infinitive in (5.33b) is unacceptable without the complementizer *for*. On the other hand, if the pronounced subject is replaced with an *un*pronounced reflexive pronoun, the result does not violate the Case Filter. The unpronounced reflexive is written *Pro*, and is the focus of extensive discussion in Chapter 6. In (5.35), its antecedent is *Sue*.

(5.35) Sue's strong preference *Pro* not to tell Pete about the decision bothers me.

Now what happens when the Obligatory Wh–Movement Constraint, the *Wh*–C Constraint and the Case Filter impose contradictory requirements on a structure? For example, suppose we consider an embedded (subordinate) *wh*–question which is infinitival and which has a pronounced subject (not *Pro*).

- If you fail to move the *wh*–phrase, the Obligatory Wh–Movement Constraint is violated, although *for* can then occupy C, satisfying the *Wh*–C Constraint and the Case Filter.
- If you do move the *wh*–phrase, then satisfying the Wh–C Constraint will require not pronouncing *for*. This will violate the Case Filter.
- If you pronounce *for*, you will satisfy the Case Filter, but violate the Wh–C Constraint.

Is this conflict resolved amiably through constraint ranking, in an OT fashion? Apparently not. No constraint seems to give way to any other. The clash among the constraints seems to lead to a situation in which no variant of the sentence can be said.

(5.36)   **Inviolable constraints?**
   a.   *Mary wonders [Bill to buy which book at the store].
        [violates Obligatory *Wh*–Movement]
   b.   *Mary wonders [which book for Bill to buy at the store].
        [violates *Wh*–C Constraint]
   c.   *Mary wonders [which book Bill to buy at the store].
        [violates Case Filter]

If the subject were the empty pronoun *Pro* (see Chapter 6) instead of *Bill*, there would be no problem, since *Pro* is not subject to the Case Filter.

(5.37)   Mary wonders [which book *Pro* to buy __ at the store].

Example (5.37) is important, since some languages (e.g. German) seem to lack infinitival questions entirely. English is not such a language, which tells us that (5.36) really does teach us something special.

Ineffability and OT

Now it would be foolish to claim that one cannot account for ineffability using the mechanics of OT. In OT, the output does not have to use the same means (words and structures) as the input, but merely has to be *maximally faithful* to the input. Deviations from FAITHFULNESS occur when some constraints outrank it. An OT account of (5.36) might, for example, rank counterparts to our two constraints above FAITHFULNESS. This would favor an output that maximally satisfies both constraints and does *not* employ the same words or structures as those found in the input. The nature of this output would depend on the details of the analysis. The output might not even receive the same semantic interpretation as that which might be associated with the input. For example, the output relevant to (5.36) might be a structure in which *Bill* is replaced with *Pro*, or *to* by *should*.

Nonetheless, an effort of this sort to handle ineffability within OT would not in and of itself constitute an argument in favor of OT constraint interaction unless one could find grounds for actually *favoring* the OT approach. Do we follow the traditional dictum "If you can't say something nice, don't say anything at all" (Clash & Crash), or the less absolutist dictum "If you can't say something nice, say the best thing you can" (OT)? One wants to know the truth.

The argument can be made stronger than this, but a serious presentation requires more background than this chapter can supply and consideration of more alternatives than I have space to present. The form of the argument is quite simple. Suppose we assume a unified FAITHFULNESS constraint. Now suppose that a particular case of ineffability can be given an OT account if some group of constraints G is ranked higher than FAITHFULNESS. This ranking, however, entails that it should always count as more important to maximally satisfy the constraints in G than for the input to match the output. If in other cases the opposite is true — i.e. it looks more important for the input to match the output than to satisfy some constraint in G — then we have discovered an argument against an OT–internal analysis of ineffability built around interactions with the unified FAITHFULNESS constraint. Now this type of argument can always be countered by a more complex OT analysis in which (as in much work on phonology) there is not a single unified FAITHFULNESS constraint, but a family of constraints of the form FAITH(X), FAITH(Y), etc. But it is at this point that one steps back to compare the complexity of the emerging OT analysis with its Clash & Crash alternative. If the complexity of the OT analysis arises precisely from the attempt to explain a Clash & Crash phenomenon with OT tools, this fact should become apparent at this point. One of the strongest arguments of this type comes from the behavior of Case in infinitival relative clauses in English. For that discussion, see Pesetsky (to appear).

In addition, Prince & Smolensky (1993) allow for situations in which the **null parse** — i.e. an unparsed candidate — is the winner of the competition among candidates. In these circumstances, presumably, some external property of language makes the unparsed candidate unusable. This, in essence, is also a Clash & Crash explanation for ineffability, since the consequence of the OT system picking the null parse while some external system rejects the null parse is ineffability.

The current literature contains interesting attempts to work out a variety of OT approaches to problems in the theory of constituency and movement, but the actual arguments favoring these accounts over alternatives in the Clash & Crash mode remain subtle and inconclusive. There is not yet any substantial body of work that both explores the interpretation of ineffability from an OT perspective and attempts to justify that perspective in light of its obvious competitor — the claim that ineffability arises when constraints clash. (There is an implicit comparison in Legendre et al. (to appear) and in Grimshaw (to appear)). Consequently, the investigation of possible OT interactions in the areas of constituency and movement is not yet a suitable topic for a general introduction such as this.[15]

A useful tactic when considering novel ideas, is to look — not for problems that have already received satisfying solutions — but for problems that are largely unsolved. This is not as easy as it sounds. One learns to live with one's unsolved problems, and with time, one becomes so used to the unsolved that it almost comes to look solved. The existence of movement imposes on the grammar a very special burden: accounting for how and where phrases are *pronounced* when they occupy more than one position. The burden actually extends to a number of cases in which unmoved phrases show interesting patterns of pronunciation and non–pronunciation. Attempts to deal with this question in the traditional program of syntactic theory have been fairly unsuccessful, and syntacticians have become used to the unsuccessful proposals that are popular.

Interestingly, it looks very likely that the systems of grammar that deal with *this* set of questions have a strongly OT character. Could this be why traditional Clash & Crash perspectives on *this* issue, unlike the others discussed above, have explained strikingly little? Let us see.

---

[15]For current collections of papers on this topic, see Barbosa et al. (to appear) and the syntax papers in Beckman et al. (1995).

> **Clash & Crash vs. OT:**
> Some laws of syntactic structure and movement seem to interact in a **Clash & Crash** fashion, rather than in an **Optimality Theoretic** manner.

3 Pronunciation of Structures with Movement

In German, Japanese and West Ulster English, we saw that the pronunciation of a moved element could be spread over more than one position. Intriguing though those examples are, they are perhaps the exception rather than the rule. For example, many (perhaps most) dialects of English that have the expression *what all* do not allow the separation of *all* from *what* in the manner of West Ulster English. Nonetheless, even when the patterns of West Ulster are unavailable, we find plentiful evidence for the full association of words and phrases with more than one position.

For example, think back on the relation between coreference and structure in examples (5.5a–b), which I reproduce as (5.38) below.

(5.38)   **Patterns of coreference and non–coreference**
    a.   He criticized the picture of John.   [*he* cannot be *John*]
    b.   His former teacher criticized the picture of John.   [*his* may be *John*]

The key factor was a prohibition on pronouns that **c–command** their antecedents. Now consider the patterns of coreference found in sentences identical to (5.38) except that the direct object has undergone *wh*–movement. Surprisingly, the judgments of coreference remain unchanged.

(5.39)   **Patterns of coreference and non–coreference are unchanged under *wh*–movement**
    a.   Which picture of John did he criticize?
        [*he* cannot be *John*]
    b.   Which picture of John did his former teacher criticize?
        [*his* may be *John*]

The facts are surprising because in neither case does the pronoun c–command the position in which its antecedent is pronounced. If "what you hear is what you get" holds for syntax, (5.39a) would be a puzzle. But if the *wh*–phrases at the front of the sentences in (5.5) are also associated with the direct object position, then the puzzle disappears. The occurrence of *which picture of John* in SPEC(CP) is not c–commanded by *he* in (5.39a), to be sure — but the occurrence of this phrase in direct object position *is* c–commanded by *he* in (5.39a) (and not c–commanded by *his* in (5.39b)).

(5.40) a. Which picture of John did he criticize ~~which picture of John~~?
            [*he* cannot be *John*]
        b. Which picture of John did his former teacher criticize ~~which picture of John~~?
            [*his* may be *John*]

Given that coreference patterns in (5.39) reveal the presence of *which picture of John* in direct object position as well as in SPEC(CP), we have to ask what principles dictate that only the higher (leftmost) occurrence of this phrase gets pronounced. In the present case, the relevant principles may seem trivial. For example, if the association of the *wh*–phrase with SPEC(CP) arises via movement from direct object position (taking seriously for the moment the "sequential" property of movement), then one might conclude that pronunciation uniquely targets the *last* position occupied by a moved phrase. Alternatively, we might note that the rightmost occurrence of the *wh*–phrase in (5.40) is structurally lower than the leftmost occurrence. An occurrence of a moved phrase that is not the highest occurrence is called a **trace**. Higher occurrences of a moved phrase are called **antecedents of the trace**. We then posit a constraint that requires a trace to be unpronounced, in effect leaving a **gap** where the trace occurs.

(5.41) SILENT TRACE: Don't pronounce the traces of a moved constituent.

The pronunciation problem does not stop here, however. As we explore this problem further, its OT flavor begins to emerge. This is nowhere clearer than in relative clause constructions, which share many properties with *wh*–questions.

Relative Clauses in English

The problem of coreference in (5.40) can be reproduced in relative clauses as well. Patterns of coreference tell us that English relative clauses, unsurprisingly, also involve *wh*–movement.

(5.42) **Patterns of coreference and non–coreference in relative clauses**
        a. Mary, whose picture of John he criticized __, is now a well–known photographer.
            [*he* cannot be *John*]
        b. Mary, whose picture of John his former teacher criticized __, is now a well–known photographer.
            [*his* may be *John*]

The data in (5.42) are straightforwardly explained if *wh*–phrases are associated with the direct object position, just as they were in (5.40). SILENT TRACE dictates that the italicized material is not pronounced.

(5.43) a. Mary, whose picture of John he criticized ~~whose picture of John~~,...
        b. Mary, whose picture of John his former teacher criticized ~~whose picture of John~~,...

In (5.42), the moved phrase is complex. The *wh*–word *whose*, which is linked to *Mary* (its antecedent), is a subpart of the phrase that undergoes *wh*–movement. Suppose we now examine simpler relative clauses in which the *wh*–word linked to the head *is* the phrase that undergoes movement. We suddenly see a greater variety of pronunciation patterns than we have seen so far. In particular, though the trace remains unpronounced, the *wh*–word need not be. Furthermore, when the *wh*–word is not pronounced, the complementizer *that* may be. In fact, the only *im*possible configuration is the one that violates the *Wh*–C Constraint (since both the *wh*–element in SPEC(CP) and the complementizer *that* are pronounced).

(5.44)   **Simple relative clauses in English: three options are available**
   a. *the person [*whom that* Mary invited __ to the party]
   b. the person [*whom ~~that~~* Mary invited __ to the party]
   c. the person [~~*whom*~~ *that* Mary invited __ to the party]
   d. the person [~~*whom that*~~ Mary invited __ to the party]

Relative clauses in which the *wh*–word is embedded in a more complex structure do not have this multiplicity of options. For example, in (5.42) the *wh*–phrase in SPEC(CP) must be pronounced. Nor can a PP in SPEC(CP) be unpronounced:[16]

(5.45)   **Complex relative clauses in English: one option**
   a. *the person [*to whom that* Mary spoke __ at the party]
   b. the person [*to whom ~~that~~* Mary spoke __ at the party]
   c. *the person [~~*to whom*~~ *that* Mary spoke __ at the party]
   d. *the person [~~*to whom that*~~ Mary spoke __ at the party]

An obvious difference between the *wh*–phrase in (5.44) and the *wh*–phrase in (5.45) is the fact that the head of the relative furnishes an antecedent for the entirety of the *wh*–phrase in (5.45), but not in (5.25). In (5.25), *person* is an antecedent for *whom*, but not for the larger phrase *to whom*.

In general, antecedentless phrases with semantic content like *to whom* in (5.45) cannot be unpronounced. This fact is accounted for by a principle called the Recoverability Condition — the idea being that the semantic content of elements that are not pronounced must be recoverable from local context. I name the corresponding OT constraint **RECOVERABILITY**; this constraint does not force pronunciation of semantically contentless words like the complementizer *that*, as in (5.46b), or of pronouns with antecedents like *whom*.

---

[16]The preposition can also be **stranded**, yielding semantically equivalent relative clauses like *the person who Mary spoke to __ at the party*. I will assume that the question of whether the preposition does or does not move to SPEC(CP) is a matter of movement, not of pronunciation. In more complex cases (infinitival relatives), one can actually argue for this claim, since any other assumption would fail to predict the optionality of stranding vs. not stranding the preposition.

(5.46)   **The complementizer in English declarative sentences**
  a. Mary thinks *that* Peter is hungry.
  b. Mary thinks ~~that~~ Peter is hungry.

The Optimality–Theoretic Character of French Relative Clauses

We have not yet seen OT in action, but we will, once we compare English relative clauses to relative clauses in French.[17] The syntax of French embedded clauses looks a lot like English. It is often particularly illuminating to study languages like English and French that differ fairly minimally. When two languages differ minimally, the differences that do exist stand out in sharper relief, and we are less likely to posit nonexistent connections among phenomena.

Interestingly, in tensed embedded clauses, the most salient differences between French and English lie in the patterns of allowable pronunciations and non–pronunciations. In particular, French is uniformly more restrictive. For example, the complementizer *que* in an embedded declarative must be pronounced, in contrast to English. That is, the alternation in (5.46) is not found in French.

(5.47)   **The complementizer in French declarative sentences**
  a. Je   crois [$_{CP}$ que Pierre a faim].
        I   believe   that Pierre is hungry.
  b. *Je crois [$_{CP}$ ~~que~~ Pierre a faim].

From a Clash & Crash perspective, one might say that French simply has a constraint that requires C to be pronounced. We will see shortly, however, that this would not be an accurate description of French.

Another obvious difference between English and French is found in relative clauses. In relative clauses with a *wh*–phrase that can be recoverably unpronounced, English allows the three patterns of pronunciation in (5.44): a **wh–relative** (with C unpronounced), a **that**–relative (with the *wh*–word unpronounced), and a **zero–relative** (with both unpronounced). In contrast, French relative clauses have only one pronunciation option: the *that* relative. All other possibilities are strongly unacceptable:[18]

---

[17]Much of what I will say about French is true of other Romance languages, such as Italian (Cinque 1981).

[18]The French data are confounded by the existence of an entirely different word *qui* (which I write *qui$_2$*), which is the shape taken by the complementizer *que* when the closest subject has undergone *wh*–movement:

(i) la table qui$_2$ est dans ma chambre
    the table which is in   my room

Qui$_2$ has properties quite unlike the *qui* discussed in the text. For example, it is compatible with the relativization of inanimate NPs, unlike the *qui* discussed in the text, which is strictly animate (and translates English *who*). Also, it appears as an alternate of *que* introducing simple declarative clauses from which the subject has undergone *wh*–movement:

(ii) Quelle table penses–tu [qui$_2$ est dans ma chambre]?
     Which table think you   that is in   my room
     'Which table do you think is in my room?'

(5.48)   **Simple relative clauses in French: one option**
  a. *l'homme [$_{CP}$ qui que je connais]   (*the man who that I know)
  b. *l'homme [$_{CP}$ qui ~~que~~ je connais]   (the man who I know)
  c. l'homme [$_{CP}$ ~~qui~~ que je connais]   (the man that I know)
  d. *l'homme [$_{CP}$ ~~qui~~ ~~que~~ je connais]   (the man I know)

What lies behind these two differences between English and French? I would like to suggest that these differences have a common origin. The problems of (5.47) and (5.48) display an obvious similarity. In each case, the only acceptable pattern of pronunciation is one which makes *que* the first pronounced word of the embedded clause (CP). Suppose, then that the patterns in (5.47) and (5.48) are due to a constraint favoring just this situation, which I call **LEFTEDGE(CP)**.[19]

(5.49)   **LEFTEDGE(CP)**: The first (leftmost) pronounced word in CP must be the complementizer.[20]

It is LEFTEDGE(CP) that favors pronunciation over non–pronunciation of *que*, and it is LEFTEDGE(CP) that favors non–pronunciation over pronunciation of relative *wh*.

Now let us ask what happens when LEFTEDGE(CP) clashes with RECOVERABILITY. Is the class resolved, in the manner of Optimality Theory, by allowing an output that violates one of these two constraints, or does the class produce a crash, i.e. no acceptable output? In fact, this interaction, unlike those we have examined so far, is resolved in a strikingly optimality–theoretic fashion. When the *wh*–phrase in SPEC(CP) of a relative clause cannot be unpronounced in conformity with recoverability, it is impossible to satisfy LEFTEDGE(CP). The system does not yield a Crash, but tolerates the violation of LEFTEDGE(CP) in favor of satisfying RECOVERABILITY. In particular, the *wh*–phrase whose pronunciation prevents CP from beginning with *que* is pronounced, with no loss

---

Additionally, *qui$_2$* can cooccur with a literary use of *qui* in the constructions discussed in the Section 5. Throughout this paper, I ignore relativization of subjects, to avoid discussing *qui$_2$*. Subject relativization in English also has special properties, which I ignore here.

[19]In Pesetsky (to appear), I argue that LEFTEDGE(CP) more generally requires the first pronounced word in a clause to be one of the "functional" words on the path from V to C — not necessarily C itself. LEFTEDGE(CP) as discussed in this chapter looks like a generalized alignment constraint (see Russell this volume). It turns out, however, that LEFTEDGE(CP) lacks the "gradient" property of alignment constraints explored in the Generalized Alignment literature. All examples that do not totally satisfy LEFTEDGE(CP) seem to be equally starred; it is no better to find the relevant sort of function word in second position within CP than to find it in tenth position. (The evidence for this point falls outside the scope of this chapter.) The difference between LEFTEDGE(CP) and alignment constraints may reflect an inadequacy of LEFTEDGE(CP) as formulated here, or it may be telling us something interesting.

[20]In main–clause declarative sentences, the complementizer is not found. That is, in French, simple sentences do not begin with *que* and in English simple sentences are not introduced by *that*. This fact may be a indication of some other constraint that overrules LEFTEDGE(CP) in main clauses, or it may be a property of the system that provides constituent structures.

of acceptability. That is, LEFTEDGE(CP) is ranked below RECOVERABILITY and is violable in classic OT fashion.

(5.50)   RECOVERABILITY » LEFTEDGE(CP)

On the other hand, something interesting does happen in these relative clauses that is not explained by RECOVERABILITY and LEFTEDGE(CP) alone: the complementizer itself is unpronounced, conforming to the observation that we earlier called the *Wh*–C Constraint. That is, only (5.51b) is possible.

(5.51)   **Complex relative clauses in French: one option**
  a. *l'homme [$_{CP}$ avec qui que j'ai dansé]
       *the man with whom that I danced
  b. l'homme   [$_{CP}$ avec qui ~~que~~ j'ai dansé]
       the man with whom I danced
  c. *l'homme [$_{CP}$ ~~avec qui~~ que j'ai dansé]
       *the man that I danced
  d. *l'homme [$_{CP}$ ~~avec qui~~ ~~que~~ j'ai dansé]
       *the man I danced

RECOVERABILITY and LeftEdge(CP) by themselves would allow both (5.51a) and (5.51b) to survive as winning candidates. The tableau below and elsewhere compares possible pronunciations of a given structure. (See the final section of this chapter for discussion.)

(5.52)   **Complex relative clauses in French: an incomplete result**

Candidates	RECOV	LE(CP)
☞ *l'homme [$_{CP}$ avec qui que j'ai dansé] the man with whom that I danced		*
☞ l'homme [$_{CP}$ avec qui ~~que~~ j'ai dansé]		*
*l'homme [$_{CP}$ ~~avec qui~~ que j'ai dansé]	*!	
*l'homme [$_{CP}$ ~~avec qui~~ ~~que~~ j'ai dansé]	*!	*

The impossibility of the first candidates in (5.51) and (5.52) is, of course, already familiar to us as the *Wh*–C Constraint. We will now try to explain this constraint. Clearly, the obligatory non–pronunciation of *que* in (5.51) is a fate that befalls *que* when other factors make it impossible for *que* to be pronounced first in its clause. In other words, there exists a constraint which (unlike LEFTEDGE(CP)) favors *non–pronunciation* of *que*. This constraint is ranked lower than LEFTEDGE(CP). Such a constraint can have an effect on output only when RECOVERABILITY prevents LEFTEDGE(CP) from being satisfied. I suggest that the constraint that forces *que* to be unpronounced in (5.51) more generally favors non–pronunciation of all closed–class function words — including complementizers, English *to* and others. Consequently, I call the constraint **TELEGRAPH**.

(5.53)   **TELEGRAPH:** Do not pronounce function words (e.g. complementizers).

Ranked high, such a constraint would produce "telegraphic" speech of the sort actually found among children in their second year (examples from Radford 1990).

(5.54)   Go nursery...Lucy go nursery. (Stevie, 25 months)

(5.55)   Where girl go? (Claire, 24 months)

About examples of this sort, Brown & Fraser (1963) commented "[T]he striking fact about the utterances of the younger children, when they are approached from the vantage point of adult grammar, is that they are almost all classifiable as grammatical sentences from which certain morphemes have been omitted." It is tempting to see this a sign that TELEGRAPH is present in the child, ranked very high.

In adult French, however, TELEGRAPH must be ranked below LEFTEDGE(CP).

(5.56)   RECOVERABILITY » LEFTEDGE(CP) » TELEGRAPH

This is not obvious from inspection of complex relative clauses alone, as (5.57) makes clear. Here, TELEGRAPH could be ranked higher than LEFTEDGE(CP), with no difference in the result.

(5.57)   **Complex relative clauses in French**

Candidates	RECOV	LE(CP)	TEL
*l'homme [$_{CP}$ avec qui que j'ai dansé] the man with whom that I danced		*	*!
☞ l'homme [$_{CP}$ avec qui ~~que~~ j'ai dansé]		*	
*l'homme [$_{CP}$ ~~avec qui~~ que j'ai dansé]	*!		*
*l'homme [$_{CP}$ ~~avec qui que~~ j'ai dansé]	*!		

What motivates the ranking LEFTEDGE(CP) » TELEGRAPH is the fact that marking the complementizer unpronounced is something you do *only* when RECOVERABILITY prevents the CP from actually starting with a pronounced complementizer. In all other cases, the CP must start with the pronounced complementizer, as we saw in both (5.47) and (5.48).

(5.58)   **Declarative *que***

Candidates	RECOV	LE(CP)	TEL
☞ Je crois [$_{CP}$ que Pierre a faim]. I believe that Pierre is hungry]			*
*Je crois [$_{CP}$ que Pierre a faim].		*!	

(5.59) **Simple relative clauses in French**

Candidates	RECOV	LE(CP)	TEL
*l'homme [CP qui que je connais] the man who that I know		*!	*
*l'homme [CP qui que je connais]		*!	
☞ l'homme [CP qui que je connais]			*
*l'homme [CP qui que je connais]		*!	

It is also worth noting that embedded questions behave just like complex relative clauses. This is because the *wh*–phrase in a question has no antecedent, and therefore cannot be recoverably unpronounced.

(5.60) **Embedded questions in French**

Candidates	RECOV	LE(CP)	TEL
*Je me demande [CP quand que Pierre arrivera]. I wonder when that Pierre will arrive		*	*!
☞ Je me demande [CP quand ~~que~~ Pierre arrivera].		*	
*Je me demande [CP ~~quand~~ que Pierre arrivera].	*!		*
*Je me demande [CP ~~quand que~~ Pierre arrivera].	*!	*	

The interaction between LEFTEDGE(CP) and TELEGRAPH not only yields the effects of what we called the *Wh*–C Constraint, but also can be said to partially *explain* it. What is explained is the environment of the effect: the fact that it is when a pronounced element precedes the complementizer within CP that something noteworthy happens. This is explained insofar as the effect arises from LEFTEDGE(CP), which has been independently motivated as an account of other phenomena. In fact, the absence of a pronounced complementizer with verb–second order in German — a central fact in the previous section — can itself be explained as a consequence of LEFTEDGE(CP) and TELEGRAPH, if the verb in second position is placed to the left of C. Unexplained, of course, is why TELEGRAPH exists in the first place. Nonetheless, evidence that TELEGRAPH does exist seems to be plentiful.

French relative clauses are the focus of this section because they are a phenomenon that suggests Optimality Theory at every turn. Consider, for example the most natural Clash & Crash treatment of the obligatory pronunciation of *que* in declarative clauses. One would posit a constraint against non–pronunciation of *que* in such a framework. This assumption would immediately run afoul of the non–pronunciation of *que* in relative clauses like (49). Consider now the most natural Clash & Crash treatment of (5.48). One might posit, for example, an obligatory rule of *wh*–deletion. This rule, in turn, would run afoul of the obligatory *non*–deletion of the *wh*–phrases in (5.51). One could always formulate the deletion rule so as to distinguish between simple and complex *wh*–phrases (just as one could always formulate a treatment of ineffability using OT constraint interaction) — but this would miss the point. Non–pronunciation is obligatory so long as recoverability is satisfied. This, of course, is a characteristic argument for OT interactions among constraints. Indeed, discussion of French relative clauses in the 1970s such as Kayne (1977) often used the phrase "deletion obligatory *up to recoverability*" to describe the patterns of pronunciation found in this construction. Sadly, researchers of the time did not look more deeply into the OT interactions they had discovered, and the topic languished for two decades.

4 French and English

As I have stressed throughout this chapter, one can use the language of Clash & Crash or OT to provide a description of most phenomena. We hope for something better — to discover the truth of the matter — how *is* our language faculty organized? But comparison of approaches to problems from different perspectives is not easy. There will rarely be a "knock–down" argument. Instead, one looks for somewhat nebulous distinctions like "the most natural proposals within framework X work better than the most natural proposals within framework Y", and one looks more generally at which framework advances the fastest and most effectively across fields of previously unsolved problems. Additionally, one hopes that evidence for a particular proposal will converge on that proposal from more than one direction. This is not a logical requirement for a true theory, but when it happens, it serves to boost one's confidence that the ideas on the table vaguely resemble the truth.

A theory like OT in which the laws of language are ranked offers the possibility that variation among languages can be attributed in some cases to differences in constraint ranking. If the sorts of laws natural to OT, when ranked and reranked, yield the set of actual grammars (and if a comparable demonstration is not possible using a Clash & Crash approach), we have a piece of converging evidence in favor of the OT approach. Few such comprehensive demonstrations exist, of course, but we can at least ask whether the signs look promising.

Suppose, for example, that TELEGRAPH were ranked higher than LEFTEDGE(CP) in some language. This sort of ranking would simply yield a language in which C is always unpronounced. Languages of this sort exist. Chinese, for example, lacks a pronounced complementizer in subordinate declarative clauses (though it has morphemes in other environments that may be complementizers; see Cheng 1991).

Tied Constraints

A more complex result would arise if TELEGRAPH and LEFTEDGE(CP) could be *tied* — i.e. if the output of LEFTEDGE(CP) and TELEGRAPH were the *union* of the outputs of the two possible rankings: i.e. the union output of LEFTEDGE(CP) » TELEGRAPH and the output of TELEGRAPH » LEFTEDGE(CP). In fact, that is exactly the case in English.[21]

---

[21]Actually, several types of interaction among constraints might merit the name *tie*. The one proposed here is the one that seems correct in general, though the arguments favoring it over its competitors lie outside the scope of this chapter.

First, in subordinate declarative clauses, the complementizer *that* may be pronounced (satisfying LEFTEDGE(CP) but violating TELEGRAPH) or it may be unpronounced (satisfying TELEGRAPH, but violating LEFTEDGE(CP)). Second, in simple relative clauses, there are the three patterns noted in (5.44). In a *that*–relative, the *wh*– is unpronounced and *that* is pronounced. This satisfies LEFTEDGE(CP) but violates TELEGRAPH. Next, in a *wh*–relative, *that* is unpronounced and *wh*– is pronounced. This satisfies TELEGRAPH but violates LEFTEDGE(CP). Finally, in a *zero*–relative, *that* is unpronounced and *wh*– is also unpronounced. This satisfies TELEGRAPH just as well as the *wh*–relative does, and violates LEFTEDGE(CP) just as badly as the *wh*–relative does.[22]

The fourth possible pattern, in which both *wh*– and *that* are pronounced, satisfies neither LEFTEDGE(CP) (since the relative clause does not begin with *that*) nor TELEGRAPH (since *that* is pronounced). It is therefore forbidden in English just as it was in French. These patterns can be represented with tableaux, but the reader has to bear in mind that the output of the tie between LEFTEDGE(CP) and TELEGRAPH is the union of the outputs of LEFTEDGE(CP) » TELEGRAPH and TELEGRAPH » LEFTEDGE(CP). In (5.61), the first candidate is the winner on the ranking LEFTEDGE(CP) » TELEGRAPH, while the second candidate is the winner on the ranking TELEGRAPH » LEFTEDGE(CP).

(5.61)   **Declarative complementizer *that***

Candidates	RECOV	LE(CP)	TEL
☞ I believe [$_{CP}$ that Peter is hungry].			*
☞ I believe [$_{CP}$ ~~that~~ Peter is hungry].		*	

In (5.62), which displays relative clauses with a simple *wh*–phrase, the third candidate (the *that*–relative) is the winner on the ranking LEFTEDGE(CP) » TELEGRAPH, while the second and fourth candidates (the two candidates in which *that* is unpronounced) are the winners on the ranking TELEGRAPH » LEFTEDGE(CP). Notice that the first candidate — the candidate that violates what we earlier called the *Wh*–C Constraint — is the winner on neither ranking. Thus we can continue to maintain that the *Wh*–C Constraint is *explained* by the OT system outlined here.

(5.62)   **Simple relative clauses in English**

Candidates	RECOV	LE(CP)	TEL
*the person [$_{CP}$ who that Mary invited t to the party]		*	*!
☞ the person [$_{CP}$ who ~~that~~ Mary invited t to the party]		*	
☞ the person [$_{CP}$ ~~who~~ that Mary invited t to the party]			*
☞ the person [$_{CP}$ ~~who that~~ Mary invited t to the party]		*	

Now let us turn to relative clauses with a complex *wh*–phrase. In English, as in French, RECOVERABILITY excludes non–pronunciation of the *wh*–phrase. This means that LEFTEDGE(CP) simply cannot be satisfied by any pattern of pronunciation that satisfies

---

[22]In a variety of contexts, including relative clauses built on a subject (*the person who/that arrived late*) and relative clauses not adjacent to their heads, the zero option is impossible. I do not discuss this additional restriction here.

RECOVERABILITY. When no candidate satisfies a constraint in OT, the constraint simply fails to have any winnowing effect on the candidate set. The union of the outputs of LEFTEDGE(CP) » TELEGRAPH and TELEGRAPH » LEFTEDGE(CP) in these cases is thus the same as the output of TELEGRAPH alone — namely, a candidate in which the *wh–* is pronounced but the complementizer *that* is not. This is the same output as in equivalent French relative clauses.

(5.63)   Complex relative clauses in English

Candidates	RECOV	LE(CP)	TEL
*the person [cp to whom that Mary spoke t at the party]		*	*!
☞ the person [cp to whom ~~that~~ Mary spoke t at the party]		*	
*the person [cp ~~to whom~~ that Mary spoke t at the party]	*!		*
*the person [cp ~~to whom that~~ Mary spoke t at the party]	*!		

Thus, a small difference between English and French — the difference between strict ranking and a tie among two constraints — accounts for a cluster of empirical differences between the two languages. If we can accumulate other results of this kind, they will constitute additional evidence for OT patterns of constraint interaction in this corner of syntax.

Pronunciation of Traces

Other phenomena connected with the pronunciation of syntactic units also participate in the system I am describing. Consider, for example, the role of SILENT TRACE in (5.41), the principle that requires that the trace of a moved constituent is not pronounced (i.e. that movement leaves a gap). SILENT TRACE outranks LEFTEDGE(CP) and TELEGRAPH in French. Otherwise, in a complex relative clause like those in (5.51), LEFTEDGE(CP) would favor a candidate in which the *trace* is pronounced instead of SPEC(CP). Such a pronunciation would be favored because it allows *que* to be pronounced first, while still satisfying RECOVERABILITY.

(5.64)   *l'homme [cp ~~avec qui~~ que j'ai dansé *avec qui*]

At this point, we should ask whether SILENT TRACE is a violable or inviolable constraint — i.e. whether all traces are gaps. In fact, not all traces are gaps. In certain circumstances, traces are pronounced, but the pronunciation is not necessarily what we might expect.

There is a class of phrases (called **islands**) which impose special restrictions on *wh–* movement and similar phenomena. In particular, an island may not contain a gap whose antecedent is not also contained by the island. For example, a coordinate structure such as *Mary and John* cannot contain a gap whose antecedent lies outside the gap.[23]

(5.65)   *This is the guy *who* I thought that [[Mary and ___]] were going to the movies].

---

[23]I use double square brackets [[...]] to delimit an island.

It is sometimes argued that movement itself cannot link a position inside an island with a position outside an island. But in fact, movement of this sort *is* possible, so long as the trace within the island is not a gap, but is actually pronounced. On the other hand (in languages like English, at least) a trace in an island is not pronounced as a full *wh*–phrase, but is instead pronounced as a *pronoun*. A pronoun of this sort (a pronunciation of a trace) is called a **resumptive pronoun**. Resumptive pronouns are especially common in conversational English.[24] They function, for example, as the sole legitimate pronunciation of trace when *wh*–movement has taken place out of a coordination, when *wh*–movement involves a subject of a clause that is itself a question, or when *wh*–movement of a noun phrase involves a relative clause within a relative clause.[25]

(5.66)   a.   *This is the guy *who* I thought that [[Mary and ___]] were going to the movies. (=(5.64))
         b.   This is the guy *who* I thought that [[Mary and *him*]] were going to the movies.

(5.67)   a.   *This is the guy *who* I wondered [[whether ___ was going to the movies]].
         b.   This is the guy *who* I wondered [[whether *he* was going to the movies]].

(5.68)   a.   *There is one worker *who* the company fired the employee [[that treated ___ badly]].
         b.   There is one worker *who* the company fired the employee [[that treated *him* badly]].

Moreover, resumptive pronouns (in English, at least) are largely *limited* to configurations involving islands. Simpler instances of *wh*–movement do not allow resumptive pronouns.

(5.69)   *This is the guy *who* I like *him*.

To summarize, we saw that island phenomena can be attributed to constraints that prohibit gaps in various contexts.[26] We then noted that resumptive pronouns — an alternative to gaps — are possible if and only if they are necessary as a way of avoiding island violations. But this is an absolutely classic OT interaction. Obviously, the constraint that *disfavors* resumptive pronouns (SILENT TRACE) is violable, and is ranked lower than the island constraints that disfavor gaps The lower–ranked constraint (as always in OT) is violated whenever that is the only way to satisfy the higher–ranked constraint.

---

[24]Resumptive pronouns are, in fact, *limited* to conversational styles of English. I return to this important fact in Section 5.
[25]The similarity of this configuration to that in (5.26) is discussed in the Section 5.
[26]The quest for a *unified* account of island phenomena has always been a lively topic of syntactic research. In this chapter, I sidestep the question of how many distinct island conditions there might be, using the vague term **islands** to refer to the collection of constraints that explain island effects. The proper analysis of island effects is tightly bound to the question, also sidestepped here, of sequential vs. non–sequential views of movement phenomena.

Furthermore, OT immediately gives us a way of understanding why these constructions show *pronouns* rather than full *wh*–phrases. If pronouns are a way of pronouncing a
wh–phrase which counts as "more silent", then SILENT TRACE is less violated by pronouncing the trace as a pronoun, than by pronouncing the trace in its full glory as *wh*–
phrase. This makes sense if pronouns are bundles of grammatical features without any
notional properties. Pronouncing a category as a pronoun means giving phonological
shape to one portion of its properties, but not another. Since OT tolerates constraint
violation, but minimizes such violation, SILENT TRACE — when it simply must be violated — will prefer a pronunciation that targets only one portion of the properties of trace
(i.e. a pronoun) over full pronunciation of the trace.

The proposal is schematized in tableau (5.70). I assume that one star is assessed against
SILENT TRACE for pronouncing *anything* in trace position, and an extra star for examples
in which the trace is fully pronounced:

(5.70)   Island violations saved by resumptive pronouns

Candidates	ISLANDS	SILTRC
*the guy who I thought that Mary and __ were going...	*!	
☞ the person who I thought that Mary and him were going ...		*
*the person who I thought that Mary and whom were going ...		**!

The idea that pronouns are a partial pronunciation of a fuller phrase resonates with
traditional ideas about pronouns, but has far–reaching consequences that I do not explore
in this chapter. I simply offer the case of resumptive pronouns as another example of OT
interactions in sentence pronunciation.

Here too, variation among dialects may be explainable as a consequence of constraint
reranking. This view might shed light on some data concerning resumptive pronouns in
the French of young children studied by Labelle (1990), to which we turn next.

Child French

Young French–speaking children display a pattern of relativization that differs from what
I have described for adults.[27] In particular, they differ in cases like (5.51), repeated as
(5.71a) below, in which a prepositional *wh*–phrase cannot be unpronounced in SPEC(CP)
due to RECOVERABILITY. I have indicated not only the unpronounced complementizers
but also the unpronounced traces. Labelle elicited relative clauses from 108 French–
speaking children, 3 to 6 years old, living in the Ottawa area. The experimental stimuli
were designed so as to elicit a variety of relative clauses. (Children were telling the
experimenter about the places they would like to put various stickers.)

Among the 1,348 relative clauses elicited by Labelle, *not one child* produced a relative
clause like those in (5.71), in which a prepositional *wh*–phrase is pronounced in
SPEC(CP), necessitating the non–pronunciation of complementizer *que*. Instead, they
employed various other strategies. In some cases, they left the prepositional *wh*–phrase

---

[27]Labelle's results may actually reflect an adult norm in the Quebec speech communities where she
did her research. This would transform our discussion from an observation about language
acquisition to one about dialect variation.

entirely unpronounced, in violation of RECOVERABILITY, but permitting the satisfaction of LEFTEDGE(CP), as in (5.72). In other cases, the prepositional *wh*-phrase was pronounced in its trace position as a resumptive pronoun, as in (5.73).[28]

(5.71)   SILENT TRACE and complex relative clauses: adult French
  a. l'homme [CP avec qui   ~~que~~ j'ai dansé   ~~avec qui~~]
     the man   with whom ~~that~~ I   danced ~~with whom~~
  b. l'homme [CP à qui   ~~que~~ le papa montre un dessin ~~à qui~~]
     the man   to whom ~~that~~ the dad shows a picture ~~to whom~~
  c. la boîte [CP dans laquelle ~~que~~ le camion rentre ~~dans laquelle~~]
     the box   in   which ~~that~~ the truck goes ~~in   which~~

(5.72)   RECOVERABILITY violations in child French
  la   boîte [CP ~~dans laquelle~~ que la   petite fille elle embarque ~~dans laquelle~~]
  the box   ~~in   which~~ ~~that~~ the little girl she goes   ~~in which~~
                                          (K 4 years, 4 months)

(5.73)   SILENT TRACE and complex relative clauses: child French
  a. l'homme [CP à qui   ~~que~~ le papa lui   montre un dessin]
     the man   to whom ~~that~~ the dad to–him shows a picture (cf. (5.71b))
                                                      (JF, 5 years, 0 months)

  b. la boîte [CP dans laquelle ~~que~~ le camion rentre dedans]
     the box   in   which ~~that~~ the truck goes in it   (cf. (5.71c))
                                                      (S, 4 years, 8 months)

Do these examples violate RECOVERABILITY by failing to pronounce the complex *wh*-phrase in SPEC(CP)? Probably not. RECOVERABILITY cares only that a phrase must be pronounced under certain circumstances. It does not care about *which position* that phrase is pronounced in, when it is associated with more than one position.[29] But the examples do violate SILENT TRACE.

Why are they allowed to violate SILENT TRACE? No islands are involved in these examples, after all. One very tempting answer involves differences in the ranking of SILENT TRACE with respect to LEFTEDGE(CP) in the adult grammar and in the children's grammar.

---

[28]There was a third pattern, in which the resumptive is a full NP duplicating the head of the relative.
  (i)   la boîte que la petite fille est debout sur la boîte
          the box that the little girl is standing on the box
This example is unexplained if relative clauses involve movement of a *wh*-element whose trace is partially pronounced in configurations where SILENT TRACE is violated. See Guasti and Shlonsky (1995) for some discussion.
[29]In some cases, preposition + pronoun is represented in French by a single word, e.g. *lui* 'to him' and *dedans* 'in it'.

If SILENT TRACE is ranked higher than LEFTEDGE(CP), it is more important to avoid resumptive pronouns than to satisfy LEFTEDGE(CP), and cases in which a prepositional phrase undergoes *wh*–movement indeed violate LEFTEDGE(CP). That is the adult pattern.

(5.74) Adult French pattern: SILENT TRACE » LEFTEDGE(CP)

l'homme à qui que le papa montre un dessin *à qui* the man to whom that the dad shows a picture *trace*	RECOV	SILENT TRACE	LE(CP)	TEL
[$_{CP}$ à qui que ... gap]			*	*!
☞ [$_{CP}$ à qui ~~que~~ ... gap]			*	
[$_{CP}$ ~~à qui~~ que ... gap]	*!			*
[$_{CP}$ ~~à qui que~~ ... gap]	*!			*
[$_{CP}$ à qui que ... P pronoun]		*!	*	*!
[$_{CP}$ à qui ~~que~~ ... P pronoun]		*!	*	
[$_{CP}$ ~~à qui~~ que ... P pronoun]		*!		*
[$_{CP}$ ~~à qui que~~ ... P pronoun]		*!	*	*

What about the children's pattern? The children's dialect seems to ensure that relative clauses start with a pronounced complementizer in *all* cases. This is just what we expect if the children's grammar, unlike the adult's, ranks SILENT TRACE *lower* than LEFTEDGE(CP). This ranking makes it more important to satisfy LEFTEDGE(CP) than to satisfy SILENT TRACE. A resumptive pronoun is the minimal violation of SILENT TRACE necessary to allow non–pronunciation of the *wh*–phrase in SPEC(CP). That is the children's pattern (5.75).[30]

Finally, if individual children allow both (5.72) and (5.73), they may be displaying a tie between RECOVERABILITY and SILENT TRACE. Example (5.72) satisfies SILENT TRACE but violates RECOVERABILITY. Example (5.73) satisfies RECOVERABILITY (if our discussion just now was correct), but violates SILENT TRACE. A tie between these constraints will allow both patterns.

These facts about child French are another demonstration of how minor differences in ranking seem to correlate with observed variation in linguistic knowledge and use. Significantly, the result depends on the idea that the constraints which govern pronunciation of structures with movement interact in an optimality–theoretic manner. The result is far from conclusive, of course — but it is a start. Among the many questions we need to understand is why the children's grammar systematically differs from the adult grammar in the way it does. Is there a default ranking of constraints? If so, is this a biologically rooted property of the system, or a response to some complex property of the

---

[30]However, the children in Labelle's study did not avoid constructions with a prepositional phrase in SPEC(CP). For example, questions elicited from the children did have this form.

(i)   Sur quoi on pèse?   [JF, 2 years, 0 months]
        on what one pushes
        'What does one push on?'

I do not attempt to explain the difference between questions and relative clauses here. It may relate to the observation that declarative main clauses are not introduced by *que*. It makes sense that LEFTEDGE(CP) would not disfavor pronunciation of prepositional phrases like *sur quoi* in SPEC(CP) if there is some independent reason for LEFTEDGE(CP) not to prevail in main clauses.

data to which the children are exposed? Answers to these questions and others have yet to be given.

(5.75)   **Child French pattern: LEFTEDGE(CP) » SILENT TRACE**

l'homme à qui que le papa montre un dessin *à qui* the man to whom that the dad shows a picture *trace*	RECOV	LE(CP)	SILENT TRACE	TEL
		*!		*!
[_CP_ à qui que ... gap]		*!		
[_CP_ à qui ~~que~~ ... gap]	*!			*
[_CP_ ~~à qui~~ que ... gap]	*!			*
[_CP_ ~~à qui que~~ ... gap]		*!	*	*!
[_CP_ à qui que ... P pronoun]		*!	*	
[_CP_ à qui ~~que~~ ... P pronoun]			*	*
☞ [_CP_ ~~à qui~~ que ... P pronoun]		*!	*	*
[_CP_ ~~à qui que~~ ... P pronoun]			*	*

5 On the Nature of Things

What bigger picture emerges from this discussion? We began with two fundamental facts about syntax: the organization of words and phrases into constituent structures, and the occasional association of these words and phrases with more than one position in structure. Clearly the grammar has systems of principles that determine constituent structure and the patterns of multi–association (movement) within it. We then observed that these two systems implicate a third system: a group of principles that determines how words and phrases are pronounced, in particular when they are associated with more than one structural position.

When the facts of system were laid out in this manner, we easily saw that the principles of this third system have an OT character. At the very beginning of this chapter, I asked whether this sort of conclusion could extend to the other two systems. We already saw some reasons to give a negative answer to this question. In the terms of Prince & Smolensky (1993), this negative answer would mean that structure and movement fall under GEN, not under EVAL. In essence, it looks as though the system of constraints governing pronunciation, though organized internally along OT lines, interacts with other parts of grammar in a Clash & Crash fashion. The winning candidate of the pronunciation system, for example, when it violates a constraint outside this system, can create a situation in which *nothing* is sayable — a situation of ineffability.

Recall, for example, what happened when the requirements of the *Wh*–C Constraint clashed with the Case Filter. The result was ineffability — a crash. As the discussion progressed, I suggested that the *Wh*–C Constraint is not an independent principle of language, but a consequence of LEFTEDGE(CP) and TELEGRAPH within the pronunciation system of syntax. Putting these two conclusions together, we can see that the pronunciation system — though its *internal* organization is optimality theoretic — must interact with the Case Filter in a Clash & Crash fashion. That is, internal affairs of some pieces of syntax might be OT, but their external affairs appear to be Clash & Crash.

Likewise, compare what happens when movement of an adverb crosses an island — a case considered in (5.26) — with what happens when movement of an NP crosses the same sort of island — a case considered in (5.68). Both examples are repeated below.[31]

(5.76)   *I'm wondering how the company will fire [any employees who treat their subordinates __].

(5.77)   a. *There is one worker who the company actually fired [the employee that treated __ badly].
          b. There is one worker who the company actually fired [the employee that treated him badly].

Island constraints prohibit a gap. Presumably the same island constraints are responsible for the ungrammaticality of both (5.76) and (5.77a). For (5.77a), in the true spirit of OT, there is a repair strategy — a minimal violation of SILENT TRACE (the resumptive pronoun) that permits the higher-ranked island constraint to be satisfied. In (5.76), however, there is no successful repair strategy. No resumptive pronoun for the adverb can save (5.76). Not even *thus* will work, in dialects that use this word as an adverbial pronoun:[32]

(5.78)   *I'm wondering *how* the company will fire any employees who treat their subordinates *thus*.

On the face of it, it looks as if any possibility that the pronunciation system offers for resumptive adverbs is ruled out by a constraint completely outside the pronunciation system. The constraint must mark as deviant the whole class of outputs from that system.

(5.79)   NO RESUMPTIVE ADVERBS: There are no resumptive adverbs.
          [Interacts with the pronunciation system in Clash & Crash fashion.]

Now an OT purist might at this point attempt to *reformulate* the Case Filter, principles governing adverbs, notions of faithfulness, and other aspects of the proposal, so as to cast an OT light on the entire set of problems. Perhaps the attempt would be successful in the areas discussed above. But there is one more category of evidence which suggests that not all is OT.

Consider the consequences of one simple and obvious fact about resumptive pronouns in English and French. True, a resumptive pronoun "saves" a structure that would otherwise violate an island constraint — but the structure with the resumptive pronoun is only marginally acceptable. (This fact is reflected in the restriction of such pronouns to casual registers of speech.) Gradient judgments fall outside of OT as a simple matter of logic. The system — as it is set up, and as it functions correctly in many domains — picks

---

[31]To make the cases more comparable, we really should use a relative clause instead of the question in (5.26), but it makes no difference to the outcome:
  (i)   *This is the way the company will fire [any employees who treat their subordinates __].
[32]Example (5.78) is fine, of course, with an irrelevant interpretation in which *how* talks about methods of *firing* (not methods of treating subordinates) and in which *thus* is not a resumptive pronoun for *how*, but a demonstrative pronoun with some other antecedent.

winners and losers. There are no silver or bronze medals in OT. Consequently, when we find silver and bronze medalists among possible pronunciations for certain structures, we can be certain that there are non–OT aspects to the interaction of linguistic constraints. From that point on, we *know* that there is a division of labor to be discovered. This chapter has made a preliminary attempt to discern what this division of labor might be.

Before concluding this chapter, it has to be noted that every aspect of this discussion is controversial. It is simply impossible in 1996 to identify a successful mainstream of work on OT interactions within syntax. Too little has been accomplished and too much is unknown. Grimshaw (to appear), for example, analyzes a variety of phenomena governing movement and structure in OT terms. Anderson (to appear) offers an approach to verb–second phenomena in a similar vein. Speas (this volume) does the same in the domain of anaphora — though I suspect that the phenomena she discusses may fall under the pronunciation system discussed here. In each case, a serious comparison of these suggestions with Clash & Crash alternatives still lies in the future. Nonetheless, even among the topics discussed in this chapter, there are some hints that some constraints on structure and movement may have an OT character after all.

For example, many non–literary registers of French allow apparent violations of our system — questions and relative clauses in which both the *wh*–phrase and the complementizer are pronounced.

(5.80)   Pronunciation of *wh*– and C in non–literary French
Je me demande *quand que* Pierre arrivera   [non–literary]
I wonder   when   that Pierre will arrive

There is some reason to think that these structures are not the counterexamples that they seem. In particular, it can be argued that these structures actually involve two distinct CPs — one containing the *wh*–phrase and an unpronounced complementizer, and another containing the pronounced complementizer. If (5.81) is the structure of (5.80), then each CP satisfies the pronunciation constraints of French. In the higher CP, RECOVERABILITY requires LEFTEDGE(CP) to be violated, so lower–ranked TELEGRAPH gets a chance to be satisfied. In the lower CP, LEFTEDGE(CP) is satisfied, though lower–ranked TELEGRAPH is violated.

(5.81)   Je me demande [CP *quand* ~~que~~ [CP *que* Pierre arrivera]]
I wonder   when   ~~that~~   that Pierre will arrive

I will not present the arguments for (5.81) here. (They involve complexities in the syntax of *quoi* 'what' that would require a lengthy excursus.) My point is different. If (5.81) is correct, then we have to wonder why double CPs are *only* found when otherwise no clause would satisfy LEFTEDGE(CP). For example, while complex relative clauses allow the structure seen in (5.81), simple relative clauses do not.

(5.82)   l'homme [CP avec qui   [CP que j'ai dansé]]   [non–literary French]
the man   with whom   that I   danced
          structure: l'homme [CP avec qui ~~que~~ [CP que j'ai dansé]]

(5.83)   *l'homme [cp que   [cp que je connais]]
          the man   that   that I know
              structure: l'homme [cp ~~qui~~ que [cp que j'ai dansé]]

If the proposed structure is correct in the first place, it looks as though the double CP construction is only possible when otherwise LEFTEDGE(CP) would not be satisfied in the lower of the two CPs. (I thank Denis Bouchard for this observation.) When LEFTEDGE(CP) can be satisfied in this CP without the recursion, the multiple CP structure is impossible. This observation smells enough like OT to make one wonder whether this phenomenon will not in the end support an extension of OT interactions beyond the domain of sentence pronunciation. Furthermore, in the domain of movement, it is often suggested that movement is avoided whenever it is not necessary in order to satisfy some constraint. Such **economy** conditions also smell like OT (a point also noted in Chapter 6), and have been incorporated into the systems of Grimshaw and others. Could it be the case that the internal workings of movement and structure–building also display OT organization — but perhaps interact with a separate pronunciation system in a Clash & Crash fashion? To these and many other questions, we do not yet know the answers.

6
Optimality Theory and Syntax: Null Pronouns and Control

Margaret Speas

In the preceding chapter, Pesetsky demonstrates the applicability of Optimality Theory to a wide variety of syntactic phenomena, but argues that constraint violability in syntax is restricted to principles governing the pronunciation of syntactic constituents. In this chapter, I explore the possibility that *all* syntactic constraints are violable.

In Section 1, I describe the current situation in syntactic theory, and consider the role that ranked violable constraints might play. Because of the fundamental differences between syntax and phonology, most notably, the fact that words have meanings but phonological segments do not, it is not always clear how the basic building blocks of OT should be instantiated in syntax. I sketch out an OT model of syntax which differs minimally from the currently most prominent syntactic theory. My goal is to make the fewest changes possible to the general model of syntax, so that I can focus on the question of whether constraints are violable.

In Section 2, I directly address the issue of the violability of syntactic constraints. For phonology, Prince & Smolensky (1993: 2) tellingly observed that constraint ranking was "found with surprising frequency in the literature, typically as a subsidiary remark in the presentation of complex constraints." One reason that OT grew so rapidly after its inception is the fact that it responded to a need in phonology which was quite clear once it was pointed out. By contrast, it has not in general been apparent that current syntactic analyses involve constraint ranking, or that violable constraints will answer any perceived need. In fact, it is not unusual for syntacticians to react to OT by saying that it might be an interesting approach to phonology, but that the past thirty years of research have shown that syntax is made up of inviolable principles. But is this really the case? A close look at the principles reveals widespread violability, frequently masked by hedges covering cases where the principles fail to apply in a straightforward way. These hedges are not restricted to principles governing sentence pronunciation; in fact, *every one* of the principles in current syntactic theory contains some sort of proviso which could be eliminated if it were stated as a violable constraint.

In sections 3 through 5, I take a look at an empirical domain that gives interesting results when approached from an OT perspective, namely the distribution and interpretation of **null pronouns** (i.e., pronouns which are unpronounced) in a variety of languages. I point out that constraint violability figures prominently in the statement of the principles governing both the distribution and the interpretation of these pronouns. I show that by elevating this violability to a central place in the grammar, we not only simplify the statements of the principles, now viewed as constraints, but also account for certain data that have eluded previous theories, and make more accurate predictions about possible language types than those made currently.

In the end, however, the basic questions about the nature of syntactic principles raised in Chapter 5 remain open. Although the pervasive constraint violations revealed in Section 2 are strong motivation for exploring an OT approach to all of syntax, it could turn out that null pronouns are licensed and interpreted solely by principles of sentence pronunciation, and as Pesetsky suggests, there is some other core of inviolable syntactic principles. The extent of constraint violability in syntax can be fruitfully investigated in the future by looking more closely at the arguments for inviolable principles found in Chapter 5, and by exploring the implications of the widespread violability revealed in Section 2 of this chapter.

1 Syntax in the Principles and Parameters Theory and in OT

Goals of Syntactic Theory

The goals of syntactic theory are to explain the ability of fluent speakers of a language to produce and understand the sentences of that language, and to recognize when sentences of that language are not well–formed. People recognize both how words can be combined to form phrases and sentences in their language, and how the words and phrases of a sentence are related to each other.

For example, English speakers recognize that (6.1a) is a fully well–formed sentence of English, but that (6.1b) is not.[1] Furthermore, they recognize that in (6.1c), the word *herself* must refer to *Susan* and cannot refer to *Mary*, whereas in (6.1d), the word *her* may refer to *Mary* but cannot refer to *Susan*.

(6.1) a. Pat ate an apple.
       b. *Pat an apple ate.
       c. Mary thinks that Susan saved some cake for herself.
       d. Mary thinks that Susan saved some cake for her.

A complete theory of syntax must, however, do more than provide a framework for analyzing the sentences of a particular language. It must also account for the syntactic

---

[1] This is not to say that (6.1b) cannot be used under certain circumstances, such as in a poem.

properties that all languages have in common and for the syntactic variation that exists among languages. (See Chapter 1.)

We have seen in previous chapters that Optimality Theory proposes that languages have in common a set of constraints which are violable, and that languages vary in the ranking that they impose upon these violable constraints. In this way, OT differs from the currently dominant view of syntax, known as **Principles and Parameters Theory** (PPT) (see Chomsky 1981, 1986, and Chapter 5), whose most recent version is called the Minimalist Program (MP) (Chomsky & Lasnik 1993, Chomsky 1995b). According to PPT, all languages share a core of *inviolable* **principles**, and differ syntactically as a result of how certain details (**parameters**) of each principle are stated.

Syntax is the study of:

how words are combined into phrases and sentences;
how the words and phrases in a sentence are related to each other.

Syntactic theory must account for:

the common syntactic properties of all languages;
how the syntactic properties of languages can vary.

Syntax in the Principles and Parameters Theory

To many people, the most striking thing about any two unrelated languages such as English and Japanese is how *different* they are. However, when we look more carefully at linguistic structure, we find that all human languages share certain essential properties. A *principle* in PPT is a statement which expresses a property shared by all languages. Principles cannot be violated, but they are not exactly absolute either: many principles contain a universal portion, and then an open *parameter*, which may take on different values in different languages. To see how this works, let's look at one of the principles which governs the order in which words appear in a sentence.

The grammatical English sentence (6.1a) differs from the ungrammatical English sentence (6.1b) in the order in which the **verb** *ate* and its **complement** *an apple* appear. If all languages were like English, we could account for that fact with an inviolable syntactic principle that says that verbs always *precede* their complements. However, while some languages, such as Yoruba, are like English in this respect, many are not. In Japanese and Hindi, for example, the verb *follows* its complements. (In the Japanese and Hindi examples, the suffixes *ga, o,* and *ne* are **case markers**, which indicate that the noun phrases with which they occur are to be understood as **subject** (*ga, ne*) or **object** complement (*o*) of the verb.)

This observation suggests that the principle governing the order of the verb and its complement should be modified to say simply that a verb and its complement form a **constituent**, labeled *VP* (for **verb phrase**; see Chapter 5), which can be considered a **projection** of the verb. In early versions of PPT, such as Chomsky (1981), this principle has two possible parameter settings: the verb may occur phrase–initially, or it may occur phrase–finally within its projection. English, Yoruba, Hindi and Japanese all obey the VP

Principle; in English and Yoruba, its parameter setting is (6.3a), whereas in Japanese and Hindi, it is (6.3b).

(6.2)   **a. English:**

S   V   O

Pat   ate   an apple.

b. Yoruba:

S   V   O

Jímò   ó   rí   ajá.

Jímò   PAST see dog

'Jímò saw a/the dog.'

c. Hindi:

S   O   V

Raam–ne seb   khaayaa.

Raam   apple   ate

'Raam ate an apple.'

d. Japanese:

S   O   V

Taroo–ga ringo–o tabeta.

Taroo   apple   ate

'Taroo ate an apple.'

(6.3)   **VP Principle:** A verb *V* and its direct object form a constituent *VP*.
Head Parameter:
   a. *V* occurs initially in *VP*.
   b. *V* occurs finally in *VP*.

In fact, the VP Principle is a special case of a more general one. Projections of nouns (*mothers of twins*), adjectives (*fond of babies*) and prepositions (*into the river*) are formed in the same way that the projections of verbs are formed. For example, *mothers of twins* can be analyzed as a noun phrase (NP) which is composed of the head noun *mothers* and the complement prepositional phrase *of twins*. In English, as in all languages, members of all **lexical categories** (e.g. verbs, nouns, adjectives, prepositions) combine with complement phrases to form larger constituents.

If the formation of NPs were governed by a principle independent of the one that governs the formation of verb phrases, each with its own parameter settings, we would expect to find many languages in which, say, the verb occurs at the beginning of VP, but the noun occurs at the end of NP. However, such languages are very rare. Rather, what we typically find is that languages are consistent across categories: if verbs are initial in the phrase, then all other lexical heads are too, and if verbs are final, then all other heads are final. This is true of the four languages illustrated above: in English and Yoruba, heads appear phrase–initially whereas in Japanese and Hindi, heads appear phrase–finally. This is illustrated in (6.4) and (6.5) with examples from English and Japanese.

Thus the VP Principle can be generalized to cover the location of the heads and complements of all phrases, with a *single* set of parameters, as in (6.6). The theory governing the universal properties of phrase structure is called **X–bar Theory.** (See also Chapter 5.)

(6.4)   **Head–Complement: heads are initial (English)**
   a. [$_{VP}$ *ate* an apple]
   b. [$_{NP}$ *mothers* of twins]
   c. [$_{AP}$ *fond* of babies]
   d. [$_{PP}$ *into* the river]

(6.5)   **Head–Complement: heads are final (Japanese)**

    a. [$_{VP}$ ringo–o *tabeta*]   'ate an apple'

        apple   ate

    b. [$_{NP}$ kaisui–no *osen*]   'pollution of sea water'

        sea water pollution

    c. [$_{PP}$ boo–*de*]   'with a stick'

        stick–with

(6.6)   **XP Principle:** A lexical category X and its complement form a constituent XP.

    Head Parameter:

    a. X occurs initially in XP.

    b. X occurs finally in XP.

The Minimalist Program (MP) removes as many stipulations from the parameters as possible. The only types of parameters that are allowed are those that involve grammatical features, such as Case, Agreement, and Tense.[2] Chomsky (1995b) proposes that there is a universal principle that features have to be **checked**, and in order for features to be checked, the phrase bearing them must occupy a **specifier** position of a phrase, which always precedes the head. In this position, the head can check the relevant features. Chomsky also assumes, following Kayne (1994), that the complement always starts out following the head so that it must move to a specifier position in order for its Case features to be checked. Finally, he proposes that languages differ only in whether such features are **strong**, in which case the phrase bearing them must move overtly to a pre–head position where the features can be checked; or **weak**, in which case the relation between the phrase that bears the features and the pre–head position for checking the features does not involve overt movement. (See Chapter 5 for further discussion of specifiers and movement.)

We can see how this theory would treat the head parameter by looking at the example of grammatical **Case**. We saw above that both Japanese and Hindi have endings on nouns to mark whether an NP is functioning as subject (**nominative case**) or as object (**accusative case**). In English, nouns do not have endings for Case, but English pronouns do show case distinctions: *she, he, I,* and *they* are nominative, while *her, him, me,* and *them* are accusative. Suppose that all NPs have Case features, but that they are strong in some languages, and weak in others. Specifically, if Case features are strong, as in Japanese and Hindi, complements would occupy a specifier position preceding the verb. If Case features are weak, as in English and Yoruba, the complement remains in post–verbal position. Under this view, the universal principles are the XP Principle and the **Satisfy** principle that all strong features must be checked overtly. The head parameter is effectively replaced by a stipulation that Case features are strong in Japanese and Hindi and weak in English and Yoruba.

---

[2]Following conventions in the literature, I capitalize the first letter in the names of these features when referring to them as abstract grammatical features, but not when using them in a non–technical sense; for example 'all NPs must have Case', but '-*o* is an accusative case marker'.

(6.7)   a. **XP Principle** (revised): A lexical category X and its complement and specifiers form a constituent *XP*.

b. **Satisfy:** Strong morpho–syntactic features must be checked overtly in a specifier position; weak features must be checked covertly in a specifier position..

Parameter: Case features are {strong, weak}

In MP, complements follow their heads and so must move to a pre–head, specifier position, in order to have their Case features checked. If those features are strong, then overt movement must take place in order to obey the inviolable principle Satisfy. If they are weak, then no overt movement has to take place, and Satisfy is obeyed vacuously.

The Principles and Parameters Theory in a Nutshell:

- Principles cannot be violated; all representations that violate a principle are ungrammatical.

- Principles can have different parametric settings, possibly restricted to differences in the strength of grammatical features.

An OT Treatment of Head–Complement Order Using Violable Constraints

One way within OT to treat the cross–linguistic variation in the relative position of the head of a phrase and its complement would be to propose a SATISFY constraint analogous to the MP principle of the same name. SATISFY can be ranked with respect to another constraint which forbids movement, and which, following Grimshaw (1996), I call STAY. STAY corresponds to the PPT principle known as Procrastinate. By ranking STAY with respect to SATISFY, we can simplify the statement of both principles; in particular, we do not need the arbitrary designation of features as strong or weak.

(6.8)   a. **Satisfy** (MP): *Strong* morpho–syntactic features must be checked *overtly* in a specifier position.

b. **SATISFY** (OT): Morpho–syntactic features must be checked in a specifier position.

(6.9)   a. **Procrastinate** (MP): Do not move *until it is necessary*.

b. **STAY** (OT): Do not move.

In English and Yoruba, STAY » SATISFY, so that complements follow their heads, whereas in Japanese and Hindi, SATISFY » STAY, so that complements precede their heads, as in the following tableau.[3]

---

[3]In this illustration, I place the preverbal complement in the specifier of VP. See Chapter 5 for other possibilities for locating the complement. I also assume that the input is an ordered sequence of words, and have omitted a number of possible candidates. I explain these matters in the next section. On the position of subjects, see Grimshaw (1996) and Samek–Ludovici (1996).

(6.10) **Tableau for English example (6.2a):**

Pat ate an apple	STAY	SATISFY
☞ Pat [$_{VP}$ ate an apple]		*
Pat [$_{VP}$ an apple ate]	*!	

(6.11) **Tableau for Japanese example (6.2d):**

Taroo–ga ringo–o tabeta	SATISFY	STAY
Taroo–ga [$_{VP}$ tabeta ringo–o]	*!	
☞ Taroo–ga [$_{VP}$ ringo–o tabeta]		*

In general, any case of cross–linguistic variation which is accounted for in PPT by different parameter settings or in MP by differences in feature strength is accounted for in OT by different rankings of violable constraints. Moreover, ranking the constraints does not simply replace PPT parameters or the MP notion of feature strength; it also allows the constraints to be stated in simplest way possible. If constraints were not violable, they would contain extra stipulations analogous to parameters or feature strength specifications.

Finally, some PPT principles may best be viewed as part of GEN. For example, if XP–Formation is part of GEN, then only candidates which obey that principle will be considered by EVAL.

The Organization of Syntax in MP and OT

We have just compared the PPT/MP and OT approaches to the problems of language universals and cross–linguistic variation. In this subsection, I outline an OT model for syntax that differs minimally from MP in overall organization.

In any theory of language, there must be a lexicon (the inventory of words and their properties) and a computational component (a statement of the ways in which sentences can be constructed from those words).[4] If we consider constructing a sentence out of a particular group of words, then those words constitute the **input**,[5] and the sentence or sentences that can be constructed out of those words the **output**. In OT, this mapping from input to output is mediated by the functions GEN and EVAL: GEN generates the set of possible outputs given a specific input, and EVAL selects the optimal candidate from among those created by GEN.

---

[4]As Chapter 4 makes clear, many words are constructed out of more elementary parts, called morphemes.

[5]Below, I propose that the input in OT should be an ordered sequence of words, but even if it is not ordered, repetitions of words must be allowed. Hence, the input cannot be considered a **set** in the technical sense of a collection of elements that occur only once.

General organization of syntax in OT

INPUT: group of words

GEN: creates candidate outputs for the input

EVAL: uses the constraint hierarchy to select the best candidate(s) for a
             given input from among the candidates produced by GEN. These
             constraints are ranked, and lower–ranked constraints may be
             violated.

In MP, the input is likewise described as a group of words,[6] which is mapped to the
output by the structure–building operations **Merge** and **Move**. These operations are constrained by principles of the sort we have already discussed, and by principles of
economy which mandate that derivations using these operations must make use of the
smallest number of steps possible.

General organization of MP syntax

SELECT a numeration: the input

MERGE and MOVE: creates the output from the input, adhering to
                         inviolable principles of UG and of economy.

The major difference between the two theories has to do with how language variation
(apart from lexical variation) is accounted for. In OT, such variation is a consequence of
the ranking of constraints, which are individually violable. In MP, it is expressed by the
different classification of morpho–syntactic features as strong and weak.

An OT Model of Syntax

Since work on syntax within OT is in its infancy, many of the questions addressed in this
section do not have one single answer that everyone working on syntax in OT agrees on.
What I do is sketch out a model that I believe to be on the right track, and note those areas where others have made different suggestions. The model I present uses OT termi-
nology (*GEN, EVAL, input, candidates*, etc.) but is in many ways the same as the model
assumed in MP. By restricting the number of changes we make in general assumptions,
we are able to focus on the central question of whether syntax contains inviolable principles or violable ranked constraints.

The input: We begin as in MP, by drawing some words out of the lexicon. In keeping
with Russell's treatment of word formation (Chapter 4), I assume that the words are
drawn from the lexicon fully inflected. This numeration is the basis of the input.

At this point, several questions arise, due to the fundamental difference between syntax
and phonology: the individual units in a syntactic representation have semantic

---

[6]To deal with the problem of repeated words mentioned in note 5, MP treats the input as a
numeration of words, which is a set of words, each of which is indexed for the number of times it
occurs.

properties, and the meaning of a sentence often depends upon its structure. This contrasts with phonology, since individual phonemes do not have semantic properties.

First, is there a direct mapping between the input and the meaning of the sentence? Some works in OT syntax (e.g. Grimshaw 1996, Samek–Ladovici, 1996) have assumed that there is. However, Legendre et al. (1995) and Keer & Bakovic (1996) have argued that there is not. The output of EVAL has to be interpreted semantically anyway, so it is redundant to include a semantic mapping in the input and then check it after EVAL. Similar considerations hold for MP. Thus, whether we are talking about OT or MP, the semantic structure is going to be mapped to the output, not to the input.

The second question is whether the input is structured, as Grimshaw assumes. I think it is clear that this is undesirable: if the input to GEN is structured, then some mechanism must exist to structure it. This would give us two different functions: GEN, which operates on structured input, and PRE–GEN, which structures the input. Since there is no reason for this duplication of effort, I assume that the input to GEN is not structured.

Given that the input is unstructured, there is still a question about whether it is ordered. For ease of exposition, I assume that the input is an ordered **string** of words. Given a particular numeration of words, any ordering of them is a possible input, just as in MP a given numeration can give rise to different sentences depending upon how MERGE applies. For example, if we select *<Mary, John, saw>* as the numeration, the input could be any of the six strings *saw Mary John, saw John Mary, Mary saw John, John saw Mary, Mary John saw,* and *John Mary saw.*[7]

GEN: The function GEN takes the input and gives as output all possible structures that can be assigned to the input. As noted above, if all languages obey the X–bar principles, then we can incorporate these principles as a restriction on GEN.

GEN plays the role in OT that Merge and Move play in MP. As in phonology, GEN can add or delete any words or features in the input, although candidates for which GEN has added or deleted elements will contain violations of FAITHFULNESS.

EVAL: GEN yields a set of candidates, which are evaluated for optimality by the function EVAL. An important property of OT is that a candidate set always has at least one optimal candidate. This means that for every ungrammatical representation, there is at least one competitor which is optimal. However, as in phonology, it is possible in some cases that the optimal candidate is the **null parse**, that is, the candidate in which none of the input words wind up included in the output. The null parse will violate the faithfulness constraint that dictates that items in the input must have a correspondent in the output, but as with all OT constraints, it is possible for a representation violating these constraints to be the optimal one.

---

[7] It is possible that inputs are unordered, with the ordering determined by constraints on the output. This would mean that a given numeration could give rise to several strings which are equally optimal, but which have different meanings, due to differences in word order. This possibility cannot be ruled out, because there is nothing in the theory I am outlining that says that all optimal candidates in a given tableau must have the same meaning.

> **An OT Model of Syntax**
>
> **INPUT:** A string of lexical items.
>
> **GEN:** Takes input, and yields as candidates all projections from the input, possibly limited to those that obey the X–bar principles.
>
> **EVAL:** Assesses candidates.
>
> **INTERPRET:** Interprets output semantically and phonetically.

2 The Violability of Inviolable Constraints

Despite the claim that all principles are absolute in PPT, once we begin to look closely at them, we find that violability is actually quite widespread. In this section, I illustrate this point with several principles, examining the empirical domain of each and showing that in each case, additional clauses are necessary to cover examples that violate the simplest statement of the principle. This violability has not been obvious because it can be written into a principle in the form of a **hedge**, a clause which extends the principle in order to cover problematic cases. Moreover, we find that it is not that just a few principles contain hedges. *Every principle of PPT* contains some hedge, some special clause to cover cases which do not obey a simple version of the principle.

After examining several instances, I list in (6.24) the most widely discussed PPT principles, along with the hedges that are necessary to make them work correctly. We find that the only thing that the hedges have in common is that they take care of cases that would otherwise be violations. There is no pattern to the exceptions which might allow us to say that linguistic principles are complex in some particular way. Rather, it looks like syntactic principles themselves are violable. First, let us consider the principle of **Full Interpretation**.

(6.12) **Full Interpretation:** There can be no superfluous symbols in an output representation.

This principle states that every element in an output representation must be fully interpretable phonetically and semantically, i.e. that output representations do not have extraneous words or symbols in them. While in general sentences do not have extraneous words or symbols in them, they sometimes *do* contain uninterpreted elements. For example, the italicized words in (6.13) do not have any independent meaning. Such a word is called an **expletive** (not to be confused with the expletives discussed in Chapter 2).

(6.13) **Expletives in English:**
   a. *It* is clear that constraints are inviolable.
   b. *There* are three cats on the porch.
   c. *It* rained all night.

Thus, Full Interpretation contains a hedge not made explicit in (6.12): expletives do not obey it. This is dealt with in PPT by saying that expletives delete right before the point at which they are semantically interpreted, whereas they survive phonetically. In other words, there actually *are* superfluous symbols in a representation, when those symbols are necessary to fulfill some other grammatical principle. This state of affairs lends itself nicely to an OT treatment, in which a FULL INTERPRETATION constraint can be violated when competing candidates violate a more highly–ranked constraint. In this case, the relevant constraint, called SUBJECT by Grimshaw & Samek–Ludovici (1995), corresponds to the PPT principle called the **Extended Projection Principle**. Sentences with expletive subjects obey SUBJECT, but violate FULL INTERPRETATION.

(6.14)   SUBJECT: Clauses must have a subject.

(6.15)   SUBJECT » FULL INTERPRETATION

It rained all night	SUBJECT	FULLINT
☞ It rained all night		*
rained all night	*!	

Conversely, FULL INTERPRETATION outranks SUBJECT in languages like Yaqui, a Uto–Aztecan language spoken in Arizona and in Sonora, Mexico, which lack expletive subjects, as in (6.16), from Escalante (1990:22).

(6.16)   yooko   yuk–ne.
        tomorrow   rain–FUT
        'It will rain tomorrow.'

Since SUBJECT, like FULL INTERPRETATION, is violable, it too must contain a hedge when formulated as the inviolable **Extended Projection Principle** in (6.17).

(6.17)   **Extended Projection Principle:** Clauses must have a subject, *unless their predicates have no arguments and the language lacks overt expletives.*

However, perhaps the Full Interpretation and Extended Projection principles are unusual in requiring elaborate hedges in order to retain their inviolability. Let us consider one of the most fundamental of all the syntactic principles of PPT, the **Case Filter**, to determine whether it can be considered both inviolable and unhedged.

As Chapter 5 showed, the Case Filter dictates that overt noun phrases may only occur where there is a **Case assigner**, such as a transitive verb, a preposition, or a Case–assigning complementizer like *for*.

(6.18)   a. I understood Sue's preference for Bill to tell Pete about the decision.
        b. *I understand Sue's preference Bill to tell Pete about the decision.

The Case Filter may be stated as in (6.19) = (5.34).

(6.19)   Case Filter: A pronounced NP must occupy a Case position.

In PPT, silent NPs are represented by the symbol *Pro*; see Chapter 5 for the rationale for considering understood but silent NPs to be syntactic entities.[8] The Case Filter, as stated in (6.19), allows Pro to occupy caseless positions, and thus explains why Pro appears largely in contexts where overt pronouns cannot appear. In (6.20), caseless positions are shown in boldface.

(6.20)   a.   Mary hopes ***Pro*** / **she* / **her* to see Bill.
        b.   Mary hired Bill after ***Pro*** / **her* / **she* interviewing him.
        c.   Mary hopes she / **Pro* will see Bill.
        d.   Mary hired Bill after she / **Pro* had interviewed him.

Notice that the statement of the Case Filter in (6.19) specifies *pronounced* NPs. In PPT, null pronouns are supposed to behave just like overt ones, except for being unpronounced. However, if we state the Case Filter in such a way that *Pro* is exempt, then we are stipulating an additional way that *Pro* differs from overt pronouns. In order to maintain that the Case Filter is inviolable, it must be assumed either that *Pro* is exempt, or that it receives a special null Case which is only compatible with null pronouns. Thus, the specification of *unpronounced* NPs in the Case Filter amounts to a hedge, which masks the fact that *Pro* sometimes does not obey it.

But perhaps Full Interpretation, Extended Projection, and the Case Filter all just happen to be subject to special provisos. Let us consider the syntactic principles which regulate coreference between pronouns and noun phrases, as in sentences like (6.21).

(6.21)   a.   Jane frightened herself.
        b.   Jane's sister frightened her.

The PPT principles which express these conditions are called Binding principles. The first, called Principle A, concerns reflexive and reciprocal pronouns like *herself, himself, myself,* and *each other*. Such a pronoun is called an **anaphor**. It has the distinctive property that it *must* corefer with another noun phrase, called its **antecedent**; for example in (6.21a), the anaphor *herself* must be understood as identifying the same individual identified by its antecedent *Jane*. It cannot refer to some other individual, say *Mary*.

The second principle, called Principle B, concerns ordinary personal pronouns like *she, her, he, him, his, I, me, my,* etc. Such a pronoun may, but does not have to, corefer with another noun phrase. If it does, we also call that noun phrase its antecedent. For example, in (6.21b), the pronoun *her* may be understood as identifying the same individual that *Jane* does, in which case *Jane* is its antecedent. However, it may also be understood as

---

[8]In PPT, it is conventional to represent the null subjects of nonfinite clauses in uppercase (*PRO*), called "big Pro", and other null pronouns in lower case (*pro*), called "small Pro" or "little Pro". The view I adopt is closer to that of Huang (1984) and Borer (1989), who do not treat the two as distinct. Therefore, I use *Pro* to refer to both "big Pro" and "small/little Pro".

referring to some other individual, say *Mary*, in which case it has no antecedent, at least not within the sentence itself.

However, the two Binding principles do more than simply describe the referential relations that anaphors and ordinary pronouns enter into. As Chapter 5 points out, syntactic structure places limits on the way in which anaphors, pronouns and noun phrases in a sentence may corefer. Two such limits are relevant here. First, in a structure in which an anaphor and its antecedent occur, if the anaphor is replaced by the corresponding ordinary pronoun, the antecedent of the anaphor generally *cannot* serve as the antecedent of the pronoun. For example, if *herself* is replaced by *her* in (6.21a), *Jane* cannot be the antecedent of that pronoun; it must refer to some other person, say *Mary*. Second, in a structure in which an ordinary pronoun and its antecedent occur, if the pronoun is replaced by the corresponding anaphor, the antecedent of the pronoun also *cannot* serve as the antecedent of the anaphor. For example, if *her* is replaced by *herself* in (6.21b), *Jane* can no longer be its antecedent; rather, the noun phrase *Jane's sister* must be.

The two Binding principles are usually stated using certain technical vocabulary, some of which has already been introduced in Chapter 5. First, we say that an anaphor or pronoun is **bound** by its antecedent if the antecedent c–commands the anaphor or pronoun. Thus, *herself* is bound by *Jane* in (6.21a), but *her* is not bound by *Jane* in (6.21b); even though *Jane* is the antecedent of *her*, *Jane* does not c–command *her*. A pronoun which is not bound is said to be **free**. Second, we say that the antecedent of an anaphor must be within its **governing category**, which is roughly the smallest noun phrase or clause potentially containing the anaphor and its antecedents. Pronouns directly contrast with anaphors; they may not be bound by any antecedent in their governing category, although they can be coreferent with noun phrases which do not bind them or are outside of the governing category. Reduced to essentials, the two binding principles may be stated as follows.

(6.22) a. **Principle A:** An anaphor must be bound in its governing category.
        b. **Principle B:** A pronoun must be free in its governing category.

These principles appear at first glance to be both inviolable and unhedged. However, both principles as just formulated may be violated, thus requiring that they be hedged in order that they may retain their inviolable status. First, the anaphor *herself* is not bound in its governing category in (6.23a), since its antecedent does not c–command it. Second, the pronoun is not free in its governing category in (6.23b). Nevertheless, both sentences are grammatical.

(6.23) a. A large portrait of herself decorated Jane's studio.
        b. Jane lost her temper.

Various ways of hedging the Binding principles to handle cases like (6.23) have been proposed. The details need not concern us. The point is that the inviolability of the binding principles is purchased at the price of complicating them.[9]

---

[9] See Pollard & Sag (1992) and Reinhart & Reuland (1993) for further discussion of this point.

It is beginning to appear that the special hedges needed for the principles of PPT are not limited to just a few isolated cases, but rather that they are *all* violable.

In (6.24), I list most of the best–known principles of PPT, along with the hedges which existing versions of these principles need in order to cover all cases. The particular formulations that I have chosen reflect the prevalent view among practitioners of PPT. Although these principles continue to be revised, the problem of violability that I illustrate here still remains.

(6.24)   Inviolable principles and their hedges

Principle	Essence	Hedge
Satisfy	All syntactic features must be satisfied...	...overtly if they are 'strong' and covertly at Logical Form if they are weak.
Full Interpretation	There can be no superfluous symbols in a representation...	...except symbols which delete before the interface level.
Extended Projection Principle	All clauses must have a subject...	...except for languages which lack overt expletives.
Case Filter	An NP must have Case...	...unless it is null.
Binding Principle A	An anaphor must be bound in its governing category...	...unless it is one of a special class of anaphors which need not be bound.
Binding Principle B	A pronoun must be free in its governing category...	...unless it occurs in an idiom like *lose her temper*.
Binding Principle C	A name must be free...	...unless it is an epithet.
X–bar Principles	Every category has a head, a specifier, and a complement...	...unless a given head takes no complement or has no features to check with its specifier.
Projection Principle	Lexical properties cannot be changed in the course of a derivation...	...unless derivational morphology can take place in the syntax.
Empty Category Principle	A trace must be properly governed...	...where 'proper government' means government by a lexical head or by a close enough antecedent.
Theta Criterion	All thematic roles must be assigned to an argument position, and all argument positions must receive a thematic role...	...except that the agent of a passive may be absorbed by the verb, and the thematic roles of nouns need not be syntactically realized.
Subjacency	Movement cannot skip potential landing sites...	...unless moving a 'D–Linked' *wh*–phrase.

The chart in (6.24) demonstrates the complications that are introduced by assuming that every principle is inviolable. If instead we allow violations to play a central role in the theory, we can remove the hedges. With the resulting simple statements of the principles, we can proceed to gain greater understanding of the character of principles of syntax, and especially their relation to the principles of other cognitive systems. Of course, as Pesetsky makes clear in Chapter 5, the nature of syntactic principles must ultimately be established through careful empirical study. However, (6.24) shows us that the inviolability of PPT principles cannot be taken for granted, and suggests that important insight can be obtained by pursuing an approach in which violations can occur.

In fact, MP now includes one area in which principles are allowed to be violated. The Economy Principles of Chomsky (1991, 1995b) are universal principles which restrict how a syntactic derivation may proceed and what the resulting syntactic representations may look like. These principles mandate that any derivation (or representation) be maximally *economical*, i.e., that they avoid long–distance movement and all but minimal structure. Each of these principles, paraphrased in (6.25) from Chomsky (1995b), can be violated exactly when it must be in order to obey some other principle, as indicated by the italicized portions.

(6.25)   Economy Principles

Least Effort	Make the *fewest number of moves possible*.
Procrastinate	Do not move overtly *unless overt movement is forced by some UG principle*.
Greed	Do not move X *unless X itself has a feature that is satisfied via that movement*.
Minimality	Movement must be to the *closest possible landing site*.
Minimize Chain Links	Long–distance dependencies must be *as short as possible*.

As with the other principles of PPT/MP, the Economy Principles are formulated so as to contain their violability within a clause of the principle itself. To put it another way, they specify the conditions under which certain general constraints against movement and long distance dependencies may be violated. For example, the fewest possible number of moves is none, so an unhedged version of Least Effort would be violated every time there is movement in syntax. Similarly, a strict version of Procrastinate is violated whenever there is overt movement, but this violation is acceptable as long as it is necessary in order to avoid violating some other principle.

Thus, the Economy Principles express restrictions on derivations or representations which may be violated exactly when some other principle *cannot* be violated. In other words, the Economy Principles are ranked below other principles, and so can be violated if a higher–ranking principle is at stake.

PPT/MP claims that the principles of UG are inviolable, but...

- all of these principles contain hedges which restrict the domain in which the principle is inviolable.
- the Minimalist Program makes violability explicit in its economy principles.

---

3 Null Pronouns and Control

Let us turn now to investigate a specific empirical domain. In all languages there are contexts in which NPs (more specifically, pronouns) may be left unpronounced, and their reference understood. There are specific syntactic conditions which govern when pronouns can be silent and how constructions containing them are understood. In PPT, the principles which govern the occurrence of such null pronouns are known as **licensing** conditions, while those which govern their interpretation are known as **identification** conditions.

Null Pronouns in English

Examples like those in (6.20), repeated here in (6.26), show that in English, null pronouns are licensed only in the subject position of **nonfinite** clauses (those which do not contain Tense). For example, the embedded clauses in (6.26a, b) are nonfinite, and permit null subjects, indicated by *Pro*, whereas the embedded clauses in (6.26c, d) are **finite** (contain Tense), and do not permit null subjects. *Pro* also cannot occur as a direct object or object of a preposition in English, as in (6.27).

(6.26) a. Mary hopes *Pro* / *she / *her to see Bill.
        b. Mary hired Bill after *Pro* / *her / *she interviewing him.
        c. Mary hopes she / *Pro will see Bill.
        d. Mary hired Bill after she / *Pro had interviewed him.

(6.27) a. *Mary promoted *Pro*.
        b. Mary promoted her.
        c. *Mary sent a letter to *Pro*.
        d. Mary sent a letter to her.

The complementarity of *Pro* and overt pronouns in English is not total, however. With certain verbs, such as *expect* and *prefer*, the subject of the nonfinite complement can be either *Pro* or an overt pronoun, and with certain others, such as *consider* and *believe*, the subject of the nonfinite complement must be overt.

(6.28) a. Mary expects *Pro* to have a good time.
        b. Mary expects her to have a good time.
        c. *Mary considers *Pro* to have good taste.
        d. Mary considers her to have good taste.

In standard PPT accounts, the licensing of *Pro* as the subject of a nonfinite clause is accounted for by the theory of **government**. This position is said to be **ungoverned**, whereas other positions are **governed**: the subject of a finite clause is governed by Inflection, the object of a verb or preposition is governed by that verb or preposition, and the subject of the nonfinite complement of a verb like *consider* is 'exceptionally' governed by that verb.[10] As Manzini (1983) and Bouchard (1983) have observed, this solution to the problem of *Pro* licensing undermines the theory of government, which is supposed to be a purely structural relation, and as such should neither be influenced by the content of the inflectional element (e.g. its finiteness), nor admit lexically determined exceptions.

The solution I adopt follows Bouchard (1983), who points out that Case Theory does distinguish finite from nonfinite Inflection, and often involves lexical secification. Bouchard proposes that *Pro* in English occurs in positions to which Case is not assigned. This allows *Pro* in the subject position of nonfinite clauses, as long as that position is not assigned Case by an exceptional verb such as *consider*.

The control principles (the principles for interpreting *Pro*) turn out to even more problematic for PPT than the licensing ones. Some sentences obey a rigid requirement that *Pro* must corefer with a c–commanding NP. For example in each of the sentences in (6.29), *Pro* can only be understood to refer to Mary. The null subject *Pro* is said to be **controlled** by its antecedent *Mary*. The concept of control is very much like that of binding: just as an anaphor generally must be bound by a c–commanding antecedent in a certain domain called its governing category, so *Pro* must generally be controlled by a c–commanding antecedent in a certain domain called its **control domain**.[11] Thus in (6.28) the controller *Mary* is written in boldface, the subscript i indicates that both *Mary* and *Pro* refer to the same individual, and square brackets mark the control domain.[12]

(6.29)   a. [**Mary**$_i$ hopes *Pro*$_i$ to promote Bill.]
      b. [Pat told **Mary**$_i$ *Pro*$_i$ to speak clearly.]
      c. [**Mary**$_i$ promised Pat *Pro*$_i$ to speak clearly.]
      d. [**Mary**$_i$ resigned after *Pro*$_i$ filing a protest.]
      e. [**Mary**$_i$ left after *Pro*$_i$ saying goodbye.]

In PPT, the interpretation of sentences like those in (6.29) is accounted for by a **Control** principle, which states that *Pro* must be controlled in its control domain. However, under certain conditions, as in (6.30), *Pro* need not or cannot be controlled. In (6.30a), the null subject of *to behave better* can be understood as being Bill, but it may also be understood as being some other person whose behavior affects Bill's reputation. For example, if Bill is a political candidate, this sentence could be used to describe a decision that his family had made about their behavior. In (6.30b), it need not be Pat who is to make the flowers, it could be her kindergarten class. Finally, in (6.30c), there is no potential antecedent at all in the sentence, and *Pro* is understood as referring to people in general.

---

[10] The licensing of *Pro* as the subject of the nonfinite complement of *expect* is accounted for by assuming that that complement may be introduced by a null complementizer, which governs its subject. If the complementizer occurs, then *Pro* is excluded, as in (6.28b). If it does not occur, then *Pro* is permitted, as in (6.28a).

[11] 'Control domain' is usually defined somewhat differently from 'governing category', the domain which is applicable to the Binding principles. Exactly what it is need not concern us for the moment.

[12] When two potential controllers appear in a control domain, as in (6.29b,c), only one of them is the actual controller, the choice being determined by the main verb (*tell* vs. *promise*).

(6.30)   a. *Pro* to behave better in public would help Bill's reputation.
         b. Pat asked how *Pro* to make flowers out of kleenex.
         c. It's fun *Pro* to dance.

In (6.30a), the only potential antecedent for *Pro* does not c–command it. In (6.30b), *Pro* has a c–commanding potential antecedent, but it is separated from it by the interrogative complementizer *how*. Finally, in (6.30c), *Pro* has no antecedent at all.

In PPT, this state of affairs was accounted for by means of the inviolable Control principle (6.31), which contains both a hedge, and a complicated definition of control domain, the latter being defined so that there is no such domain in sentences like the ones in (6.30); see Manzini (1983) and Huang (1984, 1989) for the precise definitions.[13]

(6.31)   **Control**: A null pronoun must be controlled in its control domain, *if it has one*.

Control is inviolable only because it includes the italicized hedge, which renders the principle moot when there is no control domain.

> In English, *Pro* occurs only in the subject position of certain nonfinite clauses, specifically those positions which are Caseless. When it occurs, it must be controlled, except in sentences in which there is no 'close enough' c–commanding antecedent.

Null Pronouns in Other Languages

In English, null pronouns are found only as subjects in nonfinite clauses. In fact, *Pro* can occur as the subject of nonfinite clauses in every language that has them. In some languages such as Spanish, *Pro* can also occur as the subject of finite clauses, as in (6.32). It can also occur as the direct object of verbs in some languages. This fact is discussed in Section 5, after the constraints on *Pro* in English have been introduced.

(6.32)   ***Pro* occurs as subject of both finite and nonfinite clauses in Spanish:**
         a. *Pro* vio   esa   película.
            saw:3SG that movie
            'S/he saw that movie.'
         b. Ana$_i$ contaba   con *Pro*$_i$ ganar.
            Ana counted:3SG with   to win
            'Ana counted on winning.'

> All languages allow *Pro* in the subject position of nonfinite clauses.
>
> Some languages allow *Pro* in the subject position of finite clauses and some allow *Pro* in object position.

---

[13]Manzini stated her principle as one of Binding. Huang based his principle on Manzini's, naming it the Generalized Control Rule.

4 An OT Account of Null Pronouns in English

I begin by observing that *Pro* is simply a pronoun which is phonologically null and which lacks all morpho–syntactic features such as Person, Number, Gender, and Case. Next, I observe that as a *null* pronoun, *Pro* is subject to the CONTROL constraint in (6.33).[14] Note that (6.33) lacks the hedge in (6.31): if *Pro* has no antecedent, then the CONTROL constraint is violated. Also, without the hedge, we can adopt a relatively simple definition of 'control domain', namely that it is the least clause which contains *Pro* and either Tense or an overt *wh*–phrase.

(6.33)   CONTROL: A null pronoun must be controlled in its control domain.

Next, as a pronoun, *Pro* is subject to a constraint corresponding to the Binding Principle B in (6.22b), which I rename the FREE PRONOUN principle.

(6.34)   FREE PRONOUN: A pronoun must be free in its governing category.

For the data considered here, it is not necessary to distinguish between the notions of governing category and control domain. However, I retain the terminological distinction, assuming that it may be necessary to account for other data, such as the behavior of pronouns within noun phrases. In the following examples, governing categories and control domains are marked by square brackets.

As Huang (1984, 1989) points out, if the control domain and governing category coincide, the principles corresponding to CONTROL and FREE PRONOUN conflict. If *Pro* is controlled, then it is not free, and if it is free, then it is not controlled. Huang concludes that *Pro* must be excluded from such contexts, such as direct–object position as in (6.27a), in favor of an overt pronoun, as in (6.27b). To account for this within OT, we can assume that both CONTROL and FREE PRONOUN outrank the faithfulness constraint MAX(*PRO*) defined in (6.35), which requires that input *Pro* must be matched by a corresponding output *Pro*.[15] The tableau is given in (6.36).

(6.35) MAX(*PRO*): If *Pro* occurs in the input, then its output correspondent is *Pro*.

The rankings just established also account for the fact that an overt pronoun rather than *Pro* appears in subject positions of finite clauses, as in (6.26c,d). The tableau appears in (6.37).

---

[14]Alternatively, CONTROL can be stated as a constraint not only on *Pro* but also on anaphoric (overt reflexive and reciprocal) pronouns, effectively incorporating Binding Principle A. See Burzio (1994, to appear); and note the description of *Pro* in Chapter 5 as a 'silent reflexive'. However, I do not explore this possibility here.

[15]For completeness, one should also consider cases in which output *Pro* corresponds to overt inputs, i.e. violations of DEP(*PRO*). However, such cases do not alter the conclusions drawn here, so I do not consider them. In addition, one should consider candidates in which other overt expressions, such as *herself*, *them*, *Pat*, etc. correspond to input *Pro*. Space limitations preclude our considering them here.

(6.36)   Tableau for (6.27b): CONTROL, FREE PRONOUN » MAX(*PRO*)

Mary promoted Pro	CONTROL	FREEPRN	MAX(*PRO*)
[Mary promoted *Pro*]	*!		
[*Mary*ᵢ promoted *Pro*ᵢ]		*!	
☞ [Mary promoted her]			*
[*Mary*ᵢ promoted herᵢ]		*!	*

(6.37)   Tableau for (6.26c): CONTROL, FREE PRONOUN » MAX(*PRO*)

Mary hopes Pro *will see Bill*	CONTROL	FREEPRN	MAX(*PRO*)
Mary hopes [*Pro* will see Bill]	*!		
Maryᵢ hopes [*Pro*ᵢ will see Bill]	*!		
☞ Mary hopes [she will see Bill]			*
Maryᵢ hopes [sheᵢ will see Bill]		*!	*

So far, we have said nothing about Case in relation to *Pro*. In PPT, it is necessary to state not only that overt NPs must appear in Case–marked positions (the Case Filter), but also that *Pro* is permitted to appear only in positions that are not Case–marked ('Caseless' positions). In OT, the second of these constraints is not needed, since *Pro* is excluded from Case–marked positions by the interaction of CONTROL and MAX(*PRO*). For example in (6.37), we see that since a Tensed clause counts as a control domain, the *Pro* subject of a Tensed clause (which is a Case–marked position) always violates CONTROL, which is ranked higher than MAX(*PRO*).

The counterpart to the Case Filter in OT is the CASE constraint in (6.38), which makes no reference to the phonological status of an NP. Since *Pro* has no morphosyntactic features aside from category features, it is exempt in virtue of having no case features.

(6.38)   CASE: Case–marked NPs must appear in Case positions.

If we assume that CASE outranks CONTROL and FREE PRONOUN, we correctly predict that the subject of a nonfinite clause generally must be *Pro*, because an overt NP in this position would violate CASE. Since such occurrences of *Pro* are generally controlled, it must be the case that CONTROL outranks FREE PRONOUN, as the tableau in (6.39) shows.

(6.39)   Tableau for (6.26c): CASE » CONTROL » FREE PRONOUN

Mary hopes Pro *to see Bill*	CASE	CTL	FRPRN	MAX(*PRO*)
[Mary hopes *Pro* to see Bill]		*!		
☞ [*Mary*ᵢ hopes *Pro*ᵢ to see Bill]			*	
[Mary hopes her to see Bill]	*!			*

Then, to account for the fact that the subject of nonfinite complements of verbs like *consider* cannot be *Pro*, as in (6.28c), we need only assume that those verbs obligatorily assign Case to those positions, so that the candidate with an overt pronoun in that position violates only the low–ranked MAX(*PRO*) constraint; the tableau is given in (6.40).

(6.40)   Tableau for (6.28d): Effect of Case–marking in complement of *consider*

Mary considers[+Case] Pro to have good taste	CASE	CTL	FPN	MX(PRO)
[Mary considers Pro to have good taste]		*!		
[Mary_i considers Pro_i to have good taste]			*!	
☞ [Mary considers her to have good taste]				*
[Mary_i considers her_i to have good taste]			*!	*

Next, to account for the fact that both *Pro* and overt NPs are licensed as subjects of nonfinite complements of verbs like *expect*, we may assume that Case is optionally assigned to those subjects. We may also assume that these options are reflected in the input numerations. The output when Case is not assigned is like that in (6.39): the candidate with controlled *Pro* is optimal. On the other hand, the output when case is assigned is like that in (6.40): the candidate with a free overt pronoun is optimal.

(6.41)   Tableau for (6.28a): Effect of no Case–marking in complement of *expect*

Mary expects[–Case] Pro to have a good time	CASE	CTL	FRPRN	MX(PRO)
[Mary expects Pro to have a good time]		*!		
☞ [Mary_i expects Pro_i to have a good time]			*	
[Mary expects her to have a good time]	*!			*
[Mary_i expects her_i to have a good time]	*!		*	*

(6.42)   Tableau for (6.28b): Effect of Case–marking in complement of *expect*

Mary expects[+Case] Pro to have a good time	CASE	CTL	FRPRN	MX(PRO)
[Mary expects Pro to have a good time]		*!		
[Mary_i expects Pro_i to have a good time]			*!	
☞ [Mary expects her to have a good time]				*
[Mary_i expects her_i to have a good time]			*!	*

Finally, consider the examples in (6.30), in which there is no possible antecedent for *Pro* within its control domain (i.e., no controller). In PPT, these cases require a hedge in the Control principle, as well as a complication of the definition of control domain. In OT, the prediction is that *Pro* can fail to be controlled exactly when the uncontrolled candidate is optimal. This is exactly what we find these cases. Moreover, *Pro*, being a pronoun, may have have as antecedent an NP outside its control domain, but it need not. Illustrative tableaux are given in (6.43)–(6.45).

In summary, I have shown how elevating constraint violations to a central position in syntactic theory provides a straightforward account of the nearly complementary distribution of *Pro* and overt pronouns in English, and of the fact that *Pro* must be controlled only when there is a c–commanding antecedent in its control domain. This account eliminates needless complications, allowing the constraints to be stated in their simplest and most revealing form.

(6.43)   Tableau for (6.30a): Potential antecedent does not c–command *Pro*

Pro *to behave ... Bill's reputation*	CASE	CONTROL	MAX(*PRO*)
☞ [*Pro* to behave ... Bill's reputation]		*	
☞ [*Pro*$_i$ to behave ... Bill$_i$'s reputation]		*	
[Him to behave ... Bill's reputation]	*!		*

(6.44)   Tableau for (6.30b): No antecedent within control domain

Pat wonders how Pro *to make flowers ...*	CASE	CTL	MAX(*PRO*)
☞ Pat wonders [how *Pro* to make flowers ...]		*	
☞ Pat$_i$ wonders [how *Pro*$_i$ to make flowers ...]		*	
Pat wonders [how her to make flowers ...]	*!		*

(6.45)   Tableau for (6.30c): No antecedent at all

It's fun Pro *to dance*	CASE	CONTROL	MAX(*PRO*)
☞ [It's fun *Pro* to dance]		*	
[It's fun us to dance]	*!		*

The ranking CONTROL, FREE PRONOUN » MAX(*PRO*) in English explains why *Pro* cannot occur in Case–marked positions (6.36, 6.37, 6.40, 6.42).

The ranking CASE » CONTROL, FREE PRONOUN explains why overt NPs cannot occur in Caseless positions (6.39, 6.41).

The ranking CONTROL » FREE PRONOUN explains why *Pro* is obligatorily controlled when there is a c–commanding antecedent in its control domain (6.39, 6.41).

The fact that CONTROL is violable explains why *Pro* need not be controlled if there is no c–commanding antecedent in its control domain (6.43–6.45).

5 The Cross–Linguistic Distribution of *Pro*

Although English restricts *Pro* to caseless positions, there are many languages which do not. In this section, I expand the scope of the analysis and show that a straightforward explanation of the crosslinguistic facts follows from the central OT assumptions that constraints are universal and that languages vary in the rankings they impose on those constraints.

Object Pro

In many languages, including English and Spanish, object *Pro* does not occur.[16] As (6.36) shows, the heart of the explanation for its absence is the ranking CONTROL, FREE PRONOUN » MAX(*PRO*). If the PPT principles corresponding to CONTROL and FREE PRONOUN are inviolable, as PPT requires, then the counterparts to the first two losing candidates in (6.36), repeated here as (6.46), in which the governing category and the control domain for *Pro* are identical, should be ungrammatical in every language (Huang 1984, 1989).

(6.46)   **Repetition of (6.36): CONTROL, FREE PRONOUN » MAX(*PRO*)**

	CONTROL	FREEPRN	MAX(*PRO*)
Mary promoted Pro			
[Mary promoted *Pro*]	*!		
[**Mary**$_i$ promoted *Pro*$_i$]		*!	
☞ [Mary promoted her]			*
[*Mary*$_i$ promoted her$_i$]		*!	*
Mary promoted Pat			**!

However, this prediction is incorrect. In many languages, including Thai and Korean, *uncontrolled* object *Pro* is found; see Hoonchamlong (1991), Yoon (1985). In (6.47) and (6.48), brackets delimit embedded clauses, and *C* = Complementizer.[17]

(6.47)   **Uncontrolled object *Pro*: Thai:**

Nit   bɔɔk wáà [Nuan hen *Pro*].
Nit   speak say [Nuan see   ]
'Nit said that Nuan saw him/her/them.'

(6.48)   **Uncontrolled object *Pro*: Korean:**

Chelswu–ka   [Yenghi–ka *Pro* hyeppakha–ess–ta]–ko   cwucangha–ess–ta.
Ch–NOM   [Ye–NOM   threaten–PST–DCL]–C   claim–PST–DCL
'Chelswu claimed that Yenghi threatened him/her/them.'

There are two accounts for the licensing of object *Pro* in the PPT literature, but neither of them is satisfactory. Rizzi (1986) says simply that verbs can license object *Pro* in some languages. Cole (1987) retains Huang's Generalized Control Rule (see note 10), but

---

[16]Campos (1986) points out that indefinite objects may be null in Spanish in certain discourse contexts. The restrictions on interpretation suggests that this is not *Pro*.

[17]Care needs to be taken here, because not all null objects are pronouns. A null object could also be a trace, as in the English sentence *Raw fish, I would never eat t*. Huang (1984) argues that all null objects in Mandarin are traces, and not pronouns, and that Mandarin differs from English in that Mandarin allows null topics which are capable of binding traces in object position. The authors mentioned above used Huang's tests to show that Korean and Thai do have null pronouns, and not traces, in object position.

adds a parameter whereby it fails to apply to case–marked *Pro* in some languages. Neither account provides a systematic explanation for the distribution of *Pro*.

In OT, we can turn Cole's stipulative account into an explanatory one, because in OT the fact that a certain constraint fails to be satisfied in certain languages is just what we expect. In particular, CONTROL is violable, and sentences with CONTROL violations are grammatical in just those languages in which other constraints outrank CONTROL. In Thai and Korean, those constraints are FREE PRONOUN and MAX(*PRO*). The tableau in (6.49) expresses Cole's insight that CONTROL can be violated in some languages.

(6.49)   Tableau for (6.47): FREE PRONOUN, MAX(*PRO*) » CONTROL

... *Nuan hen* Pro	FREEPRN	MAX(*PRO*)	CONTROL
☞ ... Nuan hen *Pro*			*
... Nuan₍ᵢ₎ hen *Pro*₍ᵢ₎	*!		
... Nuan hen kháw		*!	
'... Nuan saw him'			

To complete this account of the distribution of object *Pro*, we also need to consider inputs with overt pronouns instead of *Pro*. Languages which use *Pro* in object positions also use overt ones (e.g. *kháw* 'he/him' in Thai), but these generally have an emphatic meaning. To account for occurrences of overt pronouns in the output, we may simply assume that when the input includes overt pronouns, these pronouns do not incur a MAX(*PRO*) violation, so candidates with overt pronouns can be optimal.

Finally, the ranking MAX(*PRO*) » CONTROL also predicts that uncontrolled *Pro* occurs as the subject of finite clauses in Thai and Korean.[18] As the examples in (6.50) show, this prediction is borne out; the tableau in (6.51) is representative.

(6.50)   **Uncontrolled *Pro* in subject position**
     a. Thai:
       Wít bɔɔk wáá *Pro* mây rúu càk Pon.
       Wit speak say   not know Pon
       'Wit said that he/she/they do(es) not know Pon.'
     b. Korean:
       Chelswu–nun *Pro* Yenghi–lul   po–ass–ta–ko   ha–ess–ta.
       Chelswu–NOM   Yenghi–ACC see–PST–DCL–C say–PST–DCL
       'Chelswu said that he/she/they saw Yenghi.'

(6.51)   Tableau for (6.50a): FREE PRONOUN, MAX(*PRO*) » CONTROL

... Pro *mây rúu càk Pon*	FREEPRN	MAX(*PRO*)	CONTROL
☞ ... *Pro* mây rúu càk Pon			*
... kháw mây rúu càk Pon		*!	

---

[18]In addition, the ranking FREE PRONOUN » CONTROL predicts that subject *Pro* of embedded nonfinite clauses is uncontrolled if the embedding subject is in the binding domain of *Pro*, as in English. However, if the binding domain is the embedded clause, then the candidate with controlled *Pro* will be optimal.

> The occurrence of uncontrolled object *Pro* in Thai and Korean is a consequence of the ranking FREEPRONOUN, MAX(*PRO*) » CONTROL.

More on Subject *Pro*

Next we consider languages in which *Pro* is permitted as the subject of a finite clause, but not as object. Since object *Pro* is impossible, CONTROL and FREE PRONOUN must outrank MAX(*PRO*). But what licenses subject *Pro*? Jaeggli & Safir (1989) show that *Pro* subjects are licensed in languages which have either rich agreement morphology, like Spanish, or no agreement at all, like Mandarin.

Looking first at Spanish, we see that the finite verb contains complete information about the Person and Number of the subject, as well as about Tense. For example, the word *vio* 'he/she saw' in (6.32a) indicates that its subject is third–person–singular, and *vimos* 'we saw' that it is first–person–plural. This information may be analyzed as contained in a **subject agreement** element. Second, assuming that this element occurs as a constituent of IP (see Chapter 5), it c–commands the subject, and hence is in a position to control a *Pro* subject. Consequently, (6.32a) is grammatical because its *Pro* subject is controlled by the subject agreement element that appears as part of the finite verb. That is, we can represent (6.32a) as *Pro*$_i$ *vio*$_i$ *esa película*, in which *Pro* is controlled by the third–person–singular subject agreement element in *vio*. The tableau is given in (6.52).[19]

(6.52)   **Tableau for (6.32a): Control of subject of finite clause in Spanish**

Pro *vio esa película*	CONTROL	MAX(*PRO*)
Pro vio esa película	*!	
☞ *Pro*$_i$ vio$_i$ esa película		
Ella vio esa película		*!

On the other hand, Mandarin has no agreement morphology at all. To account for the fact that embedded *Pro* subjects are licensed in Mandarin, I assume that they can be controlled by a subject in a higher clause. In Section 3, I defined the control domain for the CONTROL constraint as the least clause containing Tense or a *wh*-complementizer. Suppose that Agreement, rather than Tense, is a factor in determining control domains. Then a language which lacks Agreement altogether may allow a higher subject to control *Pro*, since no Agreement element intervenes. Assuming that CONTROL and FREE PRONOUN outrank MAX(*PRO*), as in English and Spanish, and that Agreement is not a factor in determining governing categories, this predicts that (6.53a) is grammatical, but that (6.53b, c) are ungrammatical. In the tableaux in (6.54)–(6.56), control domains are indicated by square brackets, as before, and governing categories by angle brackets.

---

[19]Although the subject agreement element in Spanish can control subject *Pro*, it cannot not bind it. Only noun phrases (including *Pro*) can bind pronouns. If subject *Pro* were bound as well as controlled by the subject agreement element, then FREE PRONOUN would be violated, and the winning candidates with subject *Pro* would be rejected.

(6.53) a. Zhangsan shuo *Pro* bu renshi Lisi.
 Zhangsan say not know Lisi
 'Zhangsan$_i$ says that he$_i$ does not know Lisi.'
 b. *Pro bu renshi Lisi. (cf. *Ta bu renshi Lisi* 'He/she does not know Lisi')
 c. *Zhangsan bu renshi *Pro*. (cf. *Zhangsan bu renshi ta* 'Zhangsan does not know him/her')

(6.54) **Tableau for (6.53a): Embedded subject *Pro* is controlled**

Zhangsan shuo Pro bu renshi Lisi	CONTROL	FREEPRN	MAX(*PRO*)
[Zhangsan shuo <*Pro* bu renshi Lisi>]	*!		
☞ [Zhangsan$_i$ shuo <*Pro*$_i$ bu renshi Lisi>]			
[Zhangsan shuo <ta bu renshi Lisi>]			*!

(6.55) **Tableau for (6.53b): Main–clause subject *Pro* is disallowed**

Pro bu renshi Lisi	CONTROL	FREEPRN	MAX(*PRO*)
[<*Pro* bu renshi Lisi>]	*!		
[<*Pro*$_i$ bu renshi Lisi$_i$ >]		*!	
☞ [<Ta bu renshi Lisi >]			*

(6.56) **Tableau for (6.53c): Object *Pro* is disallowed**

Zhangsan bu renshi Pro	CONTROL	FREEPRN	MAX(*PRO*)
[<Zhangsan bu renshi *Pro* >]	*!		
[<Zhangsan$_i$ bu renshi *Pro*$_i$ >]		*!	
☞ [<Zhangsan bu renshi ta >]			*

Thus in languages in which CONTROL and FREE PRONOUN outrank MAX(*PRO*), there are two ways in which *Pro* can occur as the subject of finite clauses: first, if Agreement is rich enough to serve as a controller, as in Spanish; and second, if there is no Agreement, so that a higher NP can serve as controller, as in Mandarin.[20]

Typology of Languages According to the Distribution of Pro

Thus far, we have established that there are three types of languages with respect to the occurrence of subject and object *Pro* in finite clauses: (1) those like Thai and Korean, which allow both subject and object *Pro*; (2) those like English, which disallow both subject and object *Pro*; (3) those like Spanish and Mandarin, which allow subject *Pro* but not object *Pro*. These three types are accounted for by the rankings we have proposed for the CONTROL, FREE PRONOUN, and MAX(*PRO*) constraints. If FREE PRONOUN and MAX(*PRO*) outrank CONTROL, *Pro* is permitted as both subject and object of finite

---

[20] Huang's analysis of null objects as traces (see note 17) is disputed by Xu & Langendoen (1985) and Xu (1986), who essentially analyze all types of null elements in Mandarin as instances of *Pro*, as in the analyses of Thai and Korean presented here.

clauses. However, if CONTROL and FREE PRONOUN outrank MAX(PRO), Pro cannot occur in these positions unless the language has either rich agreement or no agreement at all, in which case Pro is licensed as subject of at least some finite clauses. However, none of the rankings we have considered licenses object Pro but disallows subject Pro in finite clauses. The situation is summarized in (6.57).

(6.57) **Types of language according to the distribution of Pro in finite clauses**

Language	Subject Pro	Object Pro	Account
Thai, Korean	yes	yes	FREEPRN, MAX(PRO) » CONTROL
English	no	no	CONTROL, FREEPRN » MAX(PRO)
Spanish, Mandarin	yes	no	CONTROL, FREEPRN » MAX(PRO); subject Pro is controlled by Agreement or by higher subject
?????	no	yes	?????

However, (6.57) does not show any language type in which FREE PRONOUN is dominated by CONTROL and MAX(PRO). Such a ranking gives rise to a language type in which object Pro can be controlled by the subject of a finite clause, as in (6.58), in which control domains are indicated by square brackets and governing categories by angle brackets. Moreover free (i.e. uncontrolled) subject Pro is licensed in such a language if MAX(PRO) outranks CONTROL, but not if CONTROL outranks MAX(PRO), as (6.59) and (6.60) show. Thus the specific ranking CONTROL » MAX(PRO) » FREE PRONOUN licenses controlled object Pro but disallows subject Pro in finite clauses.

(6.58) **The ranking CONTROL, MAX(PRO) » FREEPRN licenses object Pro**

Mary promoted Pro	CONTROL	MAX(PRO)	FREEPRN
[<Mary promoted Pro>]	*!		
☞ [<Mary$_i$ promoted Pro_i>]			*
<Mary promoted her>		*!	
<Mary$_i$ promoted her$_i$>		*!	*

(6.59) **Subject Pro is licensed in finite clauses if MAX(PRO) » CONTROL**

Pro promoted Bill	MAX(PRO)	CONTROL	FREEPRN
☞ [<Pro promoted Bill>]		*	
<She promoted Bill>	*!		

(6.60) **Subject Pro is not licensed in finite clauses if CONTROL » MAX(PRO)**

Pro promoted Bill	CONTROL	MAX(PRO)	FREEPRN
[<Pro promoted Bill>]	*!		
☞ <She promoted Bill>		*	

Our formulation of the CONTROL, FREE PRONOUN, and MAX(PRO) constraints thus predicts that there are no languages with freely referring object Pro (as in Thai and

Korean), but no subject *Pro* (as in English) in finite clauses. They do, however, permit languages both with and without subject *Pro* which have *Pro* objects interpreted as bound reflexive anaphors (though there may be additional constraints on anaphors which rule out bound object *Pro*.

Thus OT makes very specific and testable claims about the distribution and interpretation of subject and object *Pro* across the world's languages. It correctly characterizes the behavior of *Pro* in English, in which *Pro* occurs only as the subject of nonfinite clauses, but in which it may be either controlled or uncontrolled. It analyzes *Pro* in Thai and Korean as an uncontrolled pronoun which can appear in both subject and object position in finite clauses, and which differs from ordinary lexical pronouns only in that it has no phonological and morpho–syntactic features. It predicts that *Pro* subjects but not objects are allowed in a language, such as Spanish, in which they can be controlled by rich agreement elements; as well as in a language, such as Mandarin, in which there is no agreement at all, so that they can be controlled by an NP in a higher clause. It predicts the existence of languages in which *Pro* occurs in object position in finite clauses, controlled by the subject, unless independently needed constraints on anaphors happen to rule out this possibility. Finally, it predicts that no language can have uncontrolled object *Pro* without having subject *Pro* also.

6 Coda: On Rules, Constraints and Ordering

I have proposed in this paper that syntactic constraints are violable, and are ranked in each language, with the optimal candidate in a candidate set being grammatical and all other candidates being ungrammatical. Readers who have followed the progress of syntactic theory from the 1960s to the present may wonder whether OT syntax is resurrecting the old theory in which syntactic rules were ordered, and the order could vary from language to language. I would like briefly to put to rest the worry that I am pulling the stake out of the heart of a theory that was shown to be wrong. This is easily done by considering first how OT is different from the ordered rules theory, and then by reviewing the reasons why the ordered rules theory was abandoned.

Akmajian & Heny (1975:392) gave the following summary of syntactic rule ordering in English. The rules are intended to be applied to underlying structures in the order shown, to yield surface structures.

(6.61)   **Ordered rules in English syntax**

a.	Dative Movement	h.	Reflexivization	n.	Neg Placement
b.	Equi NP Deletion	i.	Extraposition	o.	Contraction
c.	Raising to Object	j.	*It* Deletion	p.	Subject–Aux Inversion
d.	Raising to Subject	k.	Number Agreement	q.	*Wh*–fronting
e.	*For* Deletion	l.	*There* Insertion	r.	Affix Hopping
f.	Passive	m.	Tag Formation	s.	*Do* Support
g.	Agent Deletion				

How does the claim that ranked principles determine the class of grammatical sentences in English differ from the claim that the application of ordered rules does so?

First, ordered rules are applied one by one in sequence, which is a very different kind of system from the one proposed in OT. In OT the constraints are hierarchically ranked, and candidates are evaluated with respect to that ranking as a whole. Second, in the ordered rules model, the rules themselves vary from language to language, while in OT the constraints are universal and only the ranking varies.

Besides, the reason that the ordered rule model was abandoned has nothing to do with any undesirable properties of rule ordering per se. It was abandoned for three reasons. First, the rules themselves were shown to be too specific to capture cross–linguistic and language–internal generalizations. For example, Raising to Subject and Reflexivization (6.61d, h) were found to obey the same locality conditions, and this was impossible to capture in a principled way in the ordered rules model. Second, in that model, different languages had to have different rules. Once the limitations on types of possible rules were sufficiently well understood, it became clear that languages could not have completely different rules. Third, it was shown that the specific restrictions (called **conditions** within that framework) on these rules ultimately allow us to eliminate the need for the sequential application of rules that this model assumes.

None of the concerns that led to the abandonment of the ordered rules model are relevant to OT. Constraints in OT are maximally general, they are universal, and they do not apply sequentially. Thus, although at first the proposal that constraints are ranked may remind us of the abandoned ordered rules theory, the two are actually very different.

Optimality Theory became the leading theory in phonology very quickly, because its architects made clear the pressing need for a principled theory of constraint ranking. In syntax at present, this need is not widely felt. In this chapter, I have shown that in fact constraint violation is widespread in syntax , being found in all of the principles of PPT. I have suggested that as in phonology, these violations should be elevated to a central place in the grammar, and I have sketched out what an OT model of syntax might look like. I looked specifically at the domain of the distribution and interpretation of null pronouns, demonstrating the existence of constraint violation within current theories and arguing that allowing languages to differ in their ranking of the constraints captures cross–linguistic variation that has posed serious problems for principle–and–parameter–based theories. Thus, I hope to have made an interesting case for a fully general application of Optimality Theory to syntax.

Afterword

The six chapters of this volume lay out basic properties of Optimality Theory and present examples to illustrate how the model works in different linguistic domains. Numerous important issues arose in these discussions, with most of them being left unresolved pending further research. This uncertainty represents the state of OT itself, which is still in its infancy. In our final comments here, we explore one of the unresolved issues which is central to OT research, and to linguistic research in general: *what is the nature of the input?* As with other aspects of linguistic theory, the nature of the input takes on new form and new significance when viewed from the OT perspective.

We do not need at this point to raise the parallel issue concerning the output. The output is easy to understand, at least intuitively: it is what people say, though as the reader will have discovered by now, it may have far more structure than meets the eye (or ear). By contrast, the input, being a hypothetical construct, is a harder concept to grasp. Nevertheless, and despite its output orientation, OT relies heavily on the input since FAITHFULNESS constraints require identity of input and output. If we do not know what the input is, we do not know how to evaluate those constraints.

We can approach the question of what is in the input by asking three questions: (i) what *must* be present in the input? (ii) what *cannot* be present in the input? and (iii) what *might* be present in the input? Broadly speaking, the answers to these questions categorize linguistic information into information that maintains distinctions, information that erases distinctions, and information that simply cannot make any distinctions.

What must be present?	information that maintains distinctions between distinct linguistic entities
What cannot be present?	information that removes distinctions between distinct linguistic entities.
What might be present?	information that cannot distinguish between distinct linguistic entities.

The Phonological Input

As an example of these three categories, let us consider the extent to which phonological features are specified in the input. Such degree of specification is a direct result of two standard assumptions: first, a grammar characterizes what a speaker knows of his or her language; and second, lexical entries contain all specific knowledge about individual morphemes. If, for instance, in some language (like American English) all low vowels are unround vowels (which allows the [+low, −round] vowels [æ] of *cat* and [a] of *hot*, but not a [+low, +round] vowel, such as British English [ɒ] of *caught*), then one effect of the grammar is to ensure that all [+low] vowels occur in the output as [−round] vowels.

Let us consider our questions about the input in light of American English low vowels.

- **What *must* be present?** Whether or not a vowel is [+low] is an idiosyncratic fact about a particular vowel: [+low], then, is an essential part of the lexical entry of a low vowel.

- **What *cannot* be present?** Whether a low vowel is [+back] ([a]) or [−back] ([æ]) is an idiosyncratic property of low vowels. Thus, [−back] *cannot* be present in the lexical entry of the vowel [a] in *hot* [hat]. Were [−back] present, the morpheme would be *hat* ([hæt] with [−back] [æ]), not the desired *hot*.

- **What *might* be present?** Whether a vowel is round or not is predictable if it is a low vowel, so neither [+round] nor [−round] are *necessary* in the representation of low vowels in English. If [−round] is present, no harm is done; if [+round] is present in the input, it must be absent in the output: the requirement (constraint) that low vowels be unrounded overrides any specifications in lexical entries.

In OT, where constraints reside only in the constraint hierarchy, the grammar of American English must have the ability to assign [−round] to [+low] vowels no matter how they are specified for [round] in the input. This is readily achieved by ranking LOWROUND, a constraint requiring that [+low] vowels be [−round], and FAITHLOW, requiring faithfulness to input [low] values, above FAITHROUND, a constraint requiring faithfulness to input [round] values, as summarized in (1).

(1) **Constraints relating [low] and [round] in American English and their ranking**
   a. LOWROUND: [+low] vowels are [+round]
   b. FAITHLOW: input [+low] vowels are [+low] in the output
   c. FAITHROUND: input [+round] vowels are [+round] in the output
   d. LOWROUND, FAITHLOW » FAITHROUND

Regardless of the representation for the [round] feature in the input, any input [+low] vowel must surface as [−round], as demonstrated in the **tableau des tableaux** in (2). Three inputs are shown: /a/, which is [+low, −round]; /ɒ/, which is [+low, +round]; and /A/, which is simply [+low]. Regardless of the input, the constraint hierarchy invariably selects [a] as the optimal output. Furthermore, in each case, the decision is made by the two higher-ranked constraints. The subordinate FAITHROUND plays no decisive role.

Thus, in English, for an input corresponding to the vowel [a], [+low] *must* be present. Conversely, [−back] *cannot* be present, for if it were present, [æ] would surface, not [a]. Finally, either [+round] or [−round] *might* be present, but it really doesn't matter whether either is present, because FAITHROUND is subordinate to LOWROUND and FAITHLOW.[1]

---

[1]What about [+back]? Is [+back] necessary in the input for the vowel [a], or is [+low] alone sufficient? There is motivation for a constraint LOWBACK: [+low] vowels are [+back]. Low vowels tend very strongly to be [+back], since tongue body lowering and backing are sympathetic gestures

(2) **Tableau des tableaux for output [a] in American English**

input	output	LowRound	FaithLow	FaithRound
/a/ ☞	[a]			
	[ɒ]	*!		
	[ə]		*!	
	[ʌ]	*!		*
/ɒ/ ☞	[a]			*
	[ɒ]	*!		
	[e]		*!	
	[ʌ]	*!		*
/ʌ/ ☞	[a]			*
	[ɒ]	*!		
	[e/ə]		*!	*
	[ʌ]	*!		*

Choosing the Correct Input

How, then, do we choose the correct input? There are two basic approaches to this question. The first is to propose some additional mechanism that makes the selection, a mechanism outside of the general principles of OT. A variety of such mechanisms are available: the best known from the OT literature is **lexicon optimization** (see also Chapters 1 and 3). Lexicon optimization examines tableaux des tableaux, such as (2), and determines the optimal input–output pairs by finding which one has the fewest highly ranked constraint violations. In our example, it is the completely faithful /a/ ↔ [a], which

---

physiologically. With such a constraint outranked by FAITHBACK in English, [æ] is able to surface (FAITHBACK ensures that an input [–back] will surface). More to the point, whether surface [a] corresponds to input [a] or input [ʌ] (i.e. [+low], with no [back] specification) is moot: in both cases, LOWBACK ensures that a back vowel [a] surfaces.

input	output	FaithBack	LowBack
/a/ ☞	[a]		
	[æ]		*!
	[ʌ]		*!
/ʌ/ ☞	[a]		
	[æ]		*!
	[ʌ]		*!
/æ/	[a]	*!	
☞	[æ]		
	[ʌ]		*!

If LOWBACK and LOWROUND outrank FAITHBACK and FAITHROUND, the result is a language in which there is only one low vowel, the back, unround [a] (e.g. Spanish).

has no violations at all. The optimal input, like all of the candidate inputs, are fully (i.e., redundantly) specified for phonological features. In other words, feature specifications which *might* be present *are* present in the input.

Another reasonable mechanism is to retain whatever is common to all inputs which correspond to a given output, and to remove what is not shared, the **minimal specification** approach. In our example, every input vowel is [+low], but values for [round] vary. Accordingly, the optimal input–output pair is /A/ ↔ [a]. With minimal specification, inputs are like underlying representations in classical generative phonology in that there is both a single correct input for each distinct output, and the optimal input is partially (i.e., nonredundantly) specified for phonological features. Feature specifications which *might* be present *are not* present in the input.

Both lexicon optimization and minimal specification follow the intuitive approach to the relation between input and output — in any domain, not just linguistics — as a **mapping** from input to output. We normally consider inputs to be underlying, or basic, and the outputs to be derived, or secondary. For example, when we add up a bunch of numbers, say 5, 6, and 7, we think of those numbers as input, and their sum, 18, as output. We don't know what the sum is until we have added up the numbers.

However, the relation between input and output does not have to be thought of that way. One can equally legitimately think of there being a mapping from output to input. For example, given the number 18, we can ask what bunches (analogous to **numerations** as defined in Chapter 6) of numbers add up to that number. Answers include {5, 6, 7}, {2, 4, 6, 6}, and {3, 3, 3, 3, 3, 3}. OT, as an output-oriented model, suggests considering the nature of the input in this same light: it does not matter which input is selected as long as the selected input results in the desired output. Consequently, any winning input–output pairs can be considered viable: /a/ ↔ [a], /ɒ/ ↔ [a], and /A/ ↔ [a]. According to this view, there need not be a unique input for a given output. Moreover, the input may be redundantly or nonredundantly specified.

This theory of the input represents more of a break with classical theories of underlying phonological representations than do minimal specification and lexicon optimization. Not only is it not necessary that there be a unique input for a given output, the input may contain elements, such as [ɒ], which appear in no output. The input is simply a set of forms which is associated with a given output by a particular constraint ranking. We call this approach the **OT perspective** since it is available under OT, but runs counter to traditional generative analysis.

Proponents of the OT perspective need not assume that every speaker of American English represents the vowel of, say, *hot* in all of the ways that the approach permits. It is consistent with this approach that different speakers represent it differently, even though they pronounce it the same. Suppose then that speaker A represents that vowel as /a/, whereas speaker B represents it as /ɒ/, and suppose also that both change their constraint rankings, so that input /ɒ/ results in output [ɒ], presumably as in British English. Then speaker A would continue to pronounce *hot* as [hat], whereas speaker B would now pronounce it as [hɒt]. Thus variability in inputs is supported if constraint reranking results in speakers' producing distinct outputs from apparently identical inputs. On the other hand, if all speakers actually posit identical inputs for a particular output, then reranking of constraints could never result in differences of this sort.

lexicon optimization	The optimal input is selected from all inputs corresponding to a single output by being the one which incurs the fewest highest-ranked constraint violations.
minimal specification	The optimal input contains all information that is common to all inputs corresponding to a single output but contains no information beyond that.
the OT perspective	Any input which results in the correct input-output pairing is a viable input.

The Role of Alternations in Constraining the Input

Morphophonemic alternation does, however, impose strict limits on how widely input representations can vary under the OT persepctive. In most dialects of English, the words *petal* and *pedal* are pronounced identically: [pʰέɾəl], where [ɾ] is a "flapped *r*", the raised *h* indicates aspiration, and the acute accent marks main stress. Given that there is no other evidence bearing on the form of the input for these words, the second consonant of both can be analyzed as corresponding to any of the inputs /t/, /tʰ/, /d/, and /ɾ/, despite the distinct orthographic symbols.

The proviso that *there is no other evidence bearing on the form of the input* is crucial, as we can see when we consider the words *metal* and *medal*, which are also pronounced identically, as [mέɾəl]. Both words (more precisely, both morphemes), however, have different pronunciations when certain suffixes are attached. When the suffix –*ic* is added to *metal*, the resulting word *metallic* is pronounced [mətʰǽlək]; but when the suffix –*ion* is added to *medal*, the resulting word *medallion* is pronounced [mədǽlyən]. There is no reason to believe that the difference in the output corresponding to the second consonant of these words is related to the difference between the –*ic* and –*ion* suffixes. Rather, the difference has to do with how the second consonants of the morphemes *metal* and *medal* may be represented in the input. In the case of *metal* it is /t/ or /tʰ/, and not /d/ or /ɾ/; whereas in the case of *medal* it is /d/, and not /t/, /tʰ/ or /ɾ/. The assumption that the input corresponding to [mέɾəl] can contain /d/, /t/, /tʰ/ or /ɾ/ can only be sustained if one is prepared to deny that the morphemes *metal* and *medal* do not occur in *metallic* and *medallion*.

The fact that *petal* and *pedal* do not have alternate forms in different morphological contexts whereas *metal* and *medal* do also has an effect on the input form of the second vowels of each of the latter pair of morphemes. In the former pair, we can assume that that vowel can be /ə/ or any short vowel such as /i/, /ɛ/, or /æ/, since any unstressed short vowel in English surfaces as [ə]. In the latter pair, however, we must assume that it is /æ/, since that is the form the vowel takes in the outputs for *metallic* and *medallion*, where this vowel is stressed. This means that the input forms for the morphemes *metal* and *medal* are restricted to /mɛtæl/ (or /mɛtʰæl/) and /mɛdæl/ respectively.

Another type of information typically included in inputs is the ordering of phonological segments. For example, the words *tack* and *cat* differ solely in the order of the two consonants, hence that order must be specified in the input. On the other hand, if we know that the vowel is first, then the order of the two consonants need not be included in the input, for only one sequencing of these consonants results in a possible English word: *act* [ækt] is an actual word of English, whereas *[ætk] is impossible, due to English constraints on admissible coda sequences. According to the OT perspective, inputs in which the consonants are ordered *k–t* and *t–k*, as well as inputs in which they are not ordered with respect to each other, are legitimate, and the question of how native speakers actually represent the sequence of segments of the word *act* is left open.[2]

Prosodic Structures

Given that phonological segments and aspects of their order must in general be specified in the input, what about other sorts of phonological structure? Consider the prosodic structure of morphemes, for example, the fact that *petal*, *pedal*, *metal*, and *medal* each have two syllables, the first of which is stressed; i.e., that it has the form of a binary trochaic foot (see Chapter 2). Should that structure be represented in the input? Assuming that this prosodic structure can be predicted from the constraint hierarchy applied to the segmental structure of each of these morphemes, this structure can be, but does not have to be, present in the input. Strict adherence to lexical optimization would require that this structure be present, maximizing redundancy; the minimal specification perspective would require that it not be present, minimizing redundancy. The OT perspective permits inputs both with and without the prosodic structure specified. If the prosodic structure is unspecified in the input, the constraint hierarchy will select only those outputs in which the correct structure appears. If the correct prosodic structure is specified in the input, no harm is done: the output is completely faithful to the input at least as far as prosodic structure is concerned. What happens if an incorrect prosodic structure is specified in the input, for example, an iambic structure?

The answer to this question is not straightforward, for it depends on the answer to a different question: are there cases in which the prosodic structure of a morpheme *must* be present in the input? This question is reminiscent of the long–debated question of whether or not stress patterns are phonemic (contrastive) in English.[3]

A clear case for stress patterns being at least marginally phonemic in English is presented by pairs of words like *Pascal* [pæskʰǽl] and *rascal* [ræskəl]. Given that *rascal* combines with the suffix *ity* resulting in *rascality* [ræskʰǽləri], we conclude that the inputs for the two words contain the segment sequences /pæskæl/ and /ræskæl/ respectively. However, from those sequences alone, one cannot determine that the output associated with the former contains two stresses, whereas the output associated with the

---

[2]On the other hand, given that both *ask* and *ax* are well–formed in standard English, we see that the order of coda consonants *s–k* and *k–s* must be specified in the input. Those speakers whose grammar contains a constraint which disallows output *sk* coda clusters (including those who pronounce *ask* and *ax* alike as [æks]) may nevertheless represent the word *ask* as containing the input *s–k* sequence.

[3]Given pairs of words like *ínsùlt* (noun) and *insúlt* (verb), it is sometimes concluded that the prosodic structure for at least certain words in English must be considered phonemic. On the other hand, it can also be maintained that the prosodic structures in these cases are predictable on the basis of other information also present in the input, such as the syntactic categorization of the words, and their morphemic composition (prefix and stem).

latter contains a single stress. Clearly, these inputs must be distinguished, so that each is correctly associated with the appropriate output. Probably the simplest way to do so is to require that the prosodic pattern of the former, at least, be present in the input, and to require that faithfulness to that pattern outrank the constraints that favor binary trochaic feet.

English must allow faithfulness to certain input stress patterns to outrank constraints favoring binary trochaic feet. What happens, then, if *metal* or *pedal* is lexically represented with an iambic foot? If left intact, the lexical iamb would produce stress only on the second syllable as in *balloon* [bəlún]: *[məthǽl]. Since faithfulness to an iamb is necessary for a word like *balloon*, any word which is lexically represented as an iamb would be so pronounced. Thus *metal*, *medal*, etc., cannot include iambic foot structure in the input. No prosodic structure is necessary, but if any is included, it must be the correct trochaic structure.

The Morphological Input

Classical generative morphology has made the same assumption that classical generative phonology made: that there is a single input for each distinct output. The OT perspective, however, leads us to expect that morphological inputs can have the same freedom of input representation that we found with phonological inputs; that is, a given output word (understood as a complex consisting of phonological, semantic, categorical, and structural information) might be associated with more than one input representation, which differ in how they represent information that might be present in the input.[4]

Blocking

Perhaps the most convincing evidence that output words can be associated with more than one input representation is provided by the phenomenon of **blocking**, which occurs when the existence of a particular word with a particular meaning prevents the expression of the same meaning by another word, typically one which is formed by a regular morphological process. For example, the existence of the word *went* 'go:Past' is said to block the existence of the word **goed* 'go:Past' in English; see Kiparsky (1982), Pinker (1995).

In many instances of blocking such as this one, the blocking word contains fewer morphemes than the blocked word. This suggests a constraint such as **MONOMORPH**, which favors the expression of particular meanings by words of one morpheme.

---

[4]Our concern here is the nature of the input to GEN for evaluation by a specific constraint hierarchy. Sometimes this input is a single morpheme, e.g. $[go]_V$ 'go'; sometimes it is a numeration of several morphemes, e.g. $\{[cat]_N$ 'cat', s 'more than one'$\}$. Our working assumption is that such numerations may be created randomly, and that EVAL selects the best possible output for each input; EVAL selects the null parse in the case of input numerations that are too ill–formed, e.g. $\{s$ 'more than one', d 'Past'$\}$. See also note 9.

(3) MONOMORPH: Words consist of one morpheme.

Blocking of *goed results if MONOMORPH outranks the constraints that require that the phonological and categorical content of the input numeration {[go]$_V$ 'go', ed 'Past'} be preserved in the output, as shown in the tableau in (4).

(4) A tableau for *went* 'go:Past'

[go]$_V$ 'go', ed 'Past'	MAX('PAST')	MONOMORPH	MAX(GO)	MAX(ED)
[[go]$_V$ ed]$_V$ 'go:Past'		*!		
☞ [went]$_V$ 'go:Past'			*	*
[go]$_V$ 'go'	*!			*

Then, since *went* is also the optimal output for the input consisting of the lexical item *went*, it follows that output *went* is associated with two distinct inputs, namely the numeration {go, ed} and the lexical item *went*.[5]

Zero Affixation

Our next example involves pairs of words which are identical phonologically, but which differ both semantically and categorically, such as the noun *nail* 'nail' and the verb *nail* 'attach with a nail'; henceforth [nail]$_N$ and [nail]$_V$, respectively. One familiar analyis of such pairs is to consider the noun form to be basic and the verb form to be derived by a **zero–affix**, which effects the appropriate categorical and semantic changes but adds no phonological content. This type of analysis parallels the minimal specification perspective in phonology. An alternative is one which does not postulate zero–affixes, but rather treats both the noun and the verb forms as separate inputs, paralleling the lexicon optimization perspective. Both types of analysis assume that there is exactly one input for each output. For the output consisting of [nail]$_V$, the minimal specification perspective postulates the input consisting of [nail]$_N$ plus a zero–affix with certain properties, whereas the lexicon optimization perspective postulates the input consisting of [nail]$_V$ itself.

According to the OT perspective, given the appropriate constraints, both inputs are possible. Thus, we may have speakers of English all of whom use the words [nail]$_N$ and [nail]$_V$ exactly alike, but who represent the input for [nail]$_V$ differently. For example, according to minimal specification, the lexicon might contain [nail]$_N$ and a zero–affix which derives certain verbs, including [nail]$_V$, from nouns; while according to lexicon optimization, it might contain [nail]$_N$ and [nail]$_V$, and no affix. From the OT perspective, however, it might contain all three items: [nail]$_N$, [nail]$_V$, and a zero–affix capable of deriving [nail]$_V$ from [nail]$_N$.

---

[5]For convenience, we may omit the categorical or semantic information associated with particular morphemes.

Ordinary Affixation and Back–Formation

Next we consider cases of ordinary affixation, in which the affix has both phonological and semantic content, as in the word *cats* 'more than one cat', which is transparently related to the morphemes *cat* 'cat' and *–s* 'more than one' (see Chapter 4). The output structures of this word and others like it are generally assumed without argument to correspond to unique inputs, such as the numeration $\{[cat]_N, s\}$. There are, however, two alternative inputs to consider, one structured, $[[cat]_N, s]_N$,[6] and the other unstructured (monomorphemic), $[cats]_N$. We argue that all three types of inputs are empirically necessary, but that each has its own particular distribution. For instance, we show that the monomorphemic input is inappropirate for transparent plurals.

Consider first an appropriate use of an unstructured monomorphemic input. This is exactly like the input $[went]_V$ 'go:Past' discussed above, and provides a means of representing irregular plurals, both ones for which there is an unrelated singular form (e.g. *people/person*) or no singular at all (e.g. $odds/*odd_N$). In each of these cases, we suggest that the input form is unstructured, e.g. $[people]_N$ 'more than one person'.

Is the unstructured form a possible input form for a word like *cats*, which is transparently related to the singular form *cat*? We believe not, for an input like $[cats]_N$ cannot characterize the relation between this and the singular. Thus, it simply does not characterize the speaker's knowledge of the language, that the plural *cats* is related to both the singular *cat* and the plural *s*. The constraint hierarchy must require that this knowledge be expressed, thereby requiring that the output *cats* correspond to a polymorphemic input. This leaves two possibilities: the structurally complex lexical item $[[cat]_N, s]_N$, and the numeration $\{[cat]_N, s\}$. From the OT perspective, it could be either. The numeration option is necessary for it reflects the ability to create a plural from a singular; without this option we would be incapable of creating plurals but could only reiterate plurals we had heard before.

The argument for structurally complex inputs is itself more complex. Consider a word like *player* 'one who plays', which, like *cats*, is transparently related to two morphemes, $[play]_V$ 'play' and *–er* 'one who V–s', where *V* is the verb stem to which the suffix attaches. As before, we can consider two possible inputs for the output *player*: the numeration $\{[play]_V, –er\}$, and the structured lexical item $[[play]_V, er]_N$, which incorporates the suffix.

The case for there being a lexically structured input $[[play]_V, er]_N$ related to the output $[[play]_V er]_N$ is quite strong for all speakers of English. The word *player* has several meanings in addition to 'one who plays', all of which are related to the meanings of $[play]_V$ and *–er*, but none of which can be fully predicted from those meanings, including 'actor', 'gambler', 'participant in an organized activity', 'member of a sports team', 'one who plays a musical instrument', and 'device for producing musical sounds'. Correlating these meanings with the lexically structured input $[[play]_V, er]_N$ gives a means of

---

[6]The notation $[[cat]_N, s]_N$ (equivalently $[s, [cat]_N]$) is intended to indicate that both morphemes $[cat]_N$ and *s* occur as constituents of the presumed lexical item *cats*, but not necessarily in that order. We assume that the correct output order is determined by constraints; see Chapter 4, and the subsection "Morpheme Order" below.

characterizing the systematic relation between *player* and the morphemes $[play]_V$ and *–er* while simultaneously characterizing the idiosyncratic semantic properties.[7]

We conclude, then, that inputs can include single unstructured lexical items made up of a single morpheme, including those, like $[went]_V$, whose meanings are normally expressed polymorphemically; single structured lexical items, like $[[play]_V, er]_N$; and numerations, like $\{[cat]_N, s\}$, but each has its specific role. Numerations express completely transparent morphology; structured lexical items express morphology whose form is transparent but whose semantics is not; and unstructured lexical items express irregular morphology.

Further confirmation that the existence of multiple inputs for a single output is a natural state of affairs is provided by the phenomenon of **back–formation**. Consider a dialect of English in which the word *burglar* 'one who steals' exists, but not *burgle* 'steal'. Some speakers of this dialect, we may presume, represent both the input and the output simply as $[burglar]_N$ 'one who steals'. Others recognize the occurrence of the *–er* affix in that word (the spelling is irrelevant; it has the right pronunciation), and represent the input as $[[burgle]_V, er]_N$ and the output as $[[burgle]_V er]_N$ 'one who steals'. However, they do not recognize $[burgle]_V$ as an independent lexical item. It is precisely these speakers whom we would expect to "discover" that word, and associate with it the meaning obtained by subtracting, as it were, the meaning of *–er* from the meaning of *burglar*. That is, they are able to extract the word *burgle* from the lexical representation of *burglar*. Having done so, they would have developed a grammar in which *burglar* is now associated with two inputs: the lexical item $[[burgle]_V, er]_N$, and the numeration $\{[burgle]_V, er\}$.[8]

We turn now to consider the categorical and phonological properties of morphemes, and consider which of these properties must be, might be, and cannot be present.

Categorical and Phonological Properties of Morphemes

The categorical properties of a morpheme tell the syntactic and/or semantic role(s) of the morpheme. For instance, a stem may be a noun or a verb, etc. Affixes "do" more: an affix may change the root's category (e.g. the adjective *happy* is a noun when paired with the suffix *-ness*, *happiness*); it may add grammatical information like tense, mood, number, case, (e.g. the addition of *-ed* to an English verb creates a past tense: *wash* vs. *washed*); it may add other types of information, such as reflexivity, reciprocity, reversive

---

[7]There is no reason to exclude the meaning 'one who plays' from also being associated with the complex lexical item. If this is done, then the output $[[play]_V er]_N$ 'one who plays' might be associated with two distinct inputs: the numeration $\{[play]_V, er\}$ and the lexical item $[[play]_V, er]_N$.
[8]Back–formation can occur whenever a form which is monomorphemic for some speakers is bimorphemic for others, and one of those morphemes does not occur as an independent word, but has the potential to do so. For those English speakers for whom the noun *destruction* is related to the input numeration $\{[destroy]_V, ion\}$, or to the bimorphemic lexical item $[[destroy]_V, ion]$, no back–formation is possible, because *destroy* already exists as an independent word. Similarly for those for whom only the monomorphemic input $[destruction]_N$ exists, no back–formation is possible. However, for those for whom the input is $[[destruct]_V, ion]$, back–formation is possible, resulting in the creation of the intransitive verb *destruct* 'break apart'.

action, diminutive, comparative (*happy* vs. *happier*), etc. (e.g. *phonology*, a noun, 'the study of language sound patterns' vs. *phonologist*, also a noun, 'one who does phonology').

This information is largely idiosyncratic, and thus is information which *must* be present and *cannot* be absent, for the most part. Questions arise about the role such information may play in constraints, but not about whether such information must be present.

Next, we consider whether phonological information is present at all in the input of affixes, arguing that at least sometimes such information cannot be omitted. Above, we discussed the so-called "zero–affixes", which give rise to morphological effects despite having no phonological content. Such effects may simply be the result of constraint interaction; they may also be a response to inputs of the type discussed above. Such morphemes, along with ones such as the Paamese reduplicative morpheme *RED* discussed in chapter 4, raise the issue of whether phonological content is necessary at all for morphemes.

We hypothesize that phonological content is necessary for at least some of the phonologically expressed morphemes, because the segments of affixes can be subject to the same constraints that hold of root segments. We would expect quite divergent behavior between affix segments and root segments were their sources distinct, i.e. if affixal segments were selected through constraint interaction, rather than being inherent in the input. This empirical result is consistent with the substantive claim that constraints are universal, not language–particular statements such as "the plural is expressed by a coronal fricative", which might account for English, but which would not be relevant for most other languages, such as Yawelmani and Paamese.

Morpheme Order

Once we accept that the phonological properties of morphemes are present in the input, we must face the question of whether the morphemes are linearly ordered in the input. As shown in Chapter 4, morpheme order can generally be determined by the definition and ranking of specific ALIGNMENT constraints, provided that certain categorial properties of morphemes are represented in the input, such as that a particular morpheme is a stem or an affix. Consequently, morpheme order generally need not be specified in the input.

Empirical support for this view comes from McCarthy's (1995) observation that it is not uncommon for phonological alternations to be sensitive to the order of segments in the input. If morphemes are unordered in the input, then the segments of one morpheme are not ordered with respect to segments of another morpheme and so the conditions for order-sensitive alternations are satisfied within a morpheme, predicting that such alternations will not take place when the relevant segments are in different morphemes.[9]

There are other implications of not ordering morphemes in the input. For example, this assumption results in a simpler account of the ungrammaticality of certain morphologically complex words. By definition, a word is ungrammatical in a language if it fails to be the output for any morphological input. Consider for example the Turkish

---

[9] On the other hand, if morphemes are ordered in the input, then those conditions can be satisfied if the input order gives rise to the appropriate segment sequence for the alternation. This does not appear to be the case.

word *ellerimizden* 'from our hands', consisting of the noun stem *el* 'hand', the suffix –*ler* 'plural', the possessive suffix –*imiz* 'our' (actually analyzed as two morphemes, –*im* 'first person' and –*iz* 'plural for personal proforms'), and the directional suffix –*den* 'from'. Given an input consisting of the numeration {*den, el, im, iz, ler*} and their categorizations, they can only be combined in the order indicated; other combinations, such as **denelimizler*, are ungrammatical in Turkish. This results simply from the interaction of particular alignment constraints for Turkish morphology without regard to the input order of the morphemes.

Now suppose that morpheme order is part of the input. Then given the input *el* + *ler* + *im* + *iz* + *den*, where each '+' indicates the concatenation of the morphemes it connects (thus fixing their order), the output *ellerimizden* is selected as before. However, *den* + *el* + *im* + *iz* + *ler* is also a possible input, to which the ungrammatical word **denelimizler* is maximally faithful. To rule it out, we must suppose that the alignment constraints which determine the correct morpheme order in Turkish outrank those faithfulness constraints, resulting in an output either in which the morphemes are reordered to *ellerimizden*, or to which morphemes are deleted or added so as to result in some other grammatical Turkish word (e.g. *elimiz* 'our hand') or the null parse. On the other hand, if we assume that the input to the morphological component is unordered, then the problem of dealing with inputs like *den* + *el* + *im* +*iz* + *ler* simply does not arise.[10]

In the subsection "Ordinary Affixation and Back–Formation" above, we pointed out that morphological inputs could be structured, while leaving information about the order of morphemes unspecified. However, there are examples, such as the English compound nouns *housework* 'work of housekeeping', and *workhouse* 'house of correction for minor offenders', which suggest that at least in some cases, morpheme order *must* be specified in the input in order to account for the difference in meaning between the two outputs. If the morphemes contained in the inputs associated with these words as outputs are unordered, then those inputs would have the same structure, namely [[*house*]$_N$, [*work*]$_N$]$_N$, one instance associated with the meaning of *housework* and the other with the meaning of *workhouse*.

However, it is not the *order* of the morphemes which accounts for the difference in meaning between the two compounds, but rather the *structural relationship* between them. In these, like in most other two–word compounds in English, the second word is the **head** and the first is its **complement**. In Chapters 5 and 6, it was pointed out that the head–complement relation in syntax can be expressed structurally, in terms of X–bar theory. Suppose we extend this idea to morphology, specifically for the analysis of compounds. If the structure of the compounds is represented as in (5), in which *NP* is the complement and the inner *N* is the head, the order of the morphemes in the output can be

---

[10]Eliminating morpheme order from the input, however, does not eliminate all problematic inputs. For example, in Turkish, the suffix *in* is a possessive suffix meaning 'your', which occurs in the same position in a Turkish word as the suffix *im* 'my' does, as in *elin* 'your hand'. Now consider the input numeration {*el, im, in*}. The maximally faithful outputs **elimin* and **elinim* are both ungrammatical (as are the corresponding English phrases **my your hand* and **your my hand*). To eliminate them, we must suppose that Turkish grammar contains a highly ranked constraint that effectively forbids combination of possessive suffixes associated with a single noun stem. Violating that constraint results either in the deletion of all but one of those suffixes, or the null parse, depending on how the constraint is stated.

determined by an alignment constraint that requires heads to be final in English compound words.

(5) **Presumed input structures for the compound nouns *housework* and *workhouse***
   a.   [[*house*]$_{NP}$, [*work*]$_N$]$_N$ 'work of housekeeping'
   b.   [[*house*]$_N$, [*work*]$_{NP}$]$_N$ 'house of correction for minor offenders'

The input structures in (5), however, cannot be correct as they stand, because they contain the category NP, which is a phrasal category, and phrasal categories do not belong to the morphological component of the grammar. Phrasal categories belong to syntax; moreover, they are assigned to output structures in the syntax, not to inputs, the inputs being numerations of words such as [*house*]$_N$ and [*work*]$_N$, which are themselves the output of morphology. That is, representations such as [*house*]$_{NP}$ in (5a) should be replaced by the syntactic mappings of input numerations, whose members all belong to categories of the morphological component. Such a mapping is expressed by syntactic tableaux, which we can represent schematically as $T(\{w_1, ..., w_n\})$, where $\{w_1, ..., w_n\}$ is a numeration of input words. For example, the structures in (5) may be revised as in (6).

(6) **Revised input structures for the compound nouns *housework* and *workhouse***
   a.   [T([*house*]$_N$), [*work*]$_N$]$_N$ 'work of housekeeping'
   b.   [[*house*]$_N$, T([*work*]$_N$)]$_N$ 'house of correction for minor offenders'

In this way, only morphologically–defined categories are present in the inputs of compound words; the appearance of syntactically–defined categories is the result of the application of syntactic evaluation to those inputs. The resulting outputs, the compound words themselves, are in turn re–input to syntax, as in the input numeration {*hate, housework, I*} which results in the syntactic output *I hate housework*.

A structure such as (6b) can also be part of an input numeration in morphology, for example {[[*house*]$_N$, T([*work*]$_N$)]$_N$, *s*}, which is uniquely associated with the output *workhouses*. There are two constraints in English which determine where the plural affix appears in plural compound nouns: one requires it to be suffixed to the head of the compound, and the other to the last word of the compound. In the case of *workhouses* both constraints are satisfied, whereas neither is satisfied by the candidate **workshouse*. In case the head is not the last word of the compound, for example *jack–in–the–box* 'kind of toy', where the plural affix appears depends on how those constraints are ranked. For those speakers for whom both *jacks–in–the–box* and *jack–in–the–boxes* are well formed, the constraints are tied.

When the head of a compound is itself morphologically complex, as in *matchmaker*, the potential for multiple inputs arises, as in (7).

(7) **Inputs that may be associated with output *matchmaker***
   a.   [[[*make*]$_V$, *er*]$_N$, T([*match*]$_N$)]
   b.   [[[*make*]$_V$, T([*match*]$_N$)]$_V$, *er*]

Presumably, the output structure associated with (7a) is [[[*match*]$_{NP}$ [[*make*]$_V$ *er*]$_N$]$_N$, whereas that associated with (7b) is [[[*match*]$_{NP}$ [*make*]$_V$]$_V$ *er*]$_N$. The latter case is analogous to *burglar*, discussed above, in which the verb stem *burgle* may be recognized as part of the input, but not as an independent word. As in that case, back–formation is possible, which may lead to the eventual creation of the compound verb *matchmake*, just as the compound verb *air–condition* has already been back–formed from the compound noun *air–conditioner*.

In summary, we have shown the following concerning input–output relations in morphology. First, the same multiplicity of potential inputs for a given output that is possible in phonology is also found in morphology. Second, the constraints which handle the multiplicity of inputs in morphology account for the phenomena of blocking and back–formation. Third, categorical and phonological information must generally be represented in morphological inputs. Fourth, morpheme order generally must not be represented in morphological inputs. Fifth, certain syntactic tableaux must be represented in some morphological inputs, at least those involving compounding.

The Syntactic Input

There have been several different conceptions of the relation between input and output in syntax in the history of generative syntax, but in all of them, it has been assumed that each output is uniquely associated with a specific input. For example, in the Minimalist Program (see Chapter 6), in which the input is considered to be simply a numeration of words, any given input corresponds to a multiplicity of outputs, each essentially a structured arrangement of that input numeration. The outputs which correspond to a given input may or may not differ in meaning. For example, putting aside the effect of Case–marking (see below), the input numeration {*loves, Marina, Mico*} is associated with the outputs *Marina loves Mico* and *Mico loves Marina* (more precisely, the structures which those strings conventionally represent), and these obviously differ in meaning. The converse, however, does not hold. Given an output, say the structure of *Marina loves Mico*, there presumably is a unique input, namely the numeration {*loves, Marina, Mico*}. Structurally ambiguous strings such as *Flying planes can be dangerous* actually represent distinct outputs, and each of these is uniquely related to a single input (possibly the same input).

In Generative Semantics on the other hand (see Huck & Goldsmith 1995 for recent discussion), syntactic inputs are considered to be representations of pure meaning, each of which is related to the various outputs that express that meaning. Thus given the input '*love*'('*Marina*', '*Mico*'), a purely semantic representation of some sort, we may have as output both *Marina loves Mico* and *Mico is loved by Marina*, assuming that the latter is synonymous with the former. However, as in the Miminalist Program, only one input is associated with a given output.

Partially Structured Inputs in Syntax

OT syntax, as it has been developed in Chapters 5 and 6 and elsewhere, is also committed to the view that the input consists minimally of a numeration of words, each of which includes phonological, morphosyntactic, and semantic information, thus excluding the possibility of a treatment of the input as "pure semantics". Nevertheless, the idea that the input–output relation in syntax is constrained by semantics is an attractive one, so one would like to know whether it is possible within OT to develop a theory of that relation in which semantically divergent outputs such as *Marina loves Mico* and *Mico loves Marina* are associated with distinct inputs.

For these particular sentences, a mechanism is available, namely the marking of noun phrases for Case, which may be accomplished by requiring that at least their heads be specified for Case in the input, so that, for example, *Marina* marked for nominative case ($Marina_{NOM}$) is a distinct lexical item from *Marina* marked for accusative case ($Marina_{ACC}$). Accordingly, the input numeration for *Marina loves Mico* is {*loves*, $Marina_{NOM}$, $Mico_{ACC}$}, whereas the input numeration for *Mico loves Marina* is {*loves*, $Marina_{ACC}$, $Mico_{NOM}$}. However, this mechanism does not succeed in distinguishing inputs for another class of sentences, those whose outputs differ in the location of noun phrases with the same case. For example, both *Marina* and *Mico* are specified for nominative case in the sentences *Marina said that Mico left* and *Mico said that Marina left*.

To distinguish between these outputs in input numerations, we would have to add information to the lexical items *Marina* and *Mico* about whether they can occur in a main or in a subordinate clause, e.g. $Marina_{MAIN}$ vs. $Marina_{SUB}$. But even this desperate expedient would fail to distinguish, for example, *Dante denied that Marina said that Mico left* from *Dante denied that Mico said that Marina left*, in which *Marina* and *Mico* occur in different subordinate clauses.

What to do? More information is needed about inputs than is provided by numerations of words in order to insure that semantically distinct outputs correspond to distinct inputs. The obvious answer is to structure the input. However, this obvious answer has an equally obvious objection: it is the role of constraints to evaluate the structures that can be associated with inputs. By structuring the input, the constraints are left with nothing to do except to check for faithfulness violations of input structures. A subtler answer is based on the observation that syntactic structure is one of the things that *may* be in the syntactic input. Given the OT perspective, we are free to structure the input if doing so has desirable consequences, such as providing a unique output for a particular input. One side–effect is that a given output may be associated with more than one input, but as we have seen, that already occurs under the OT perspective in both phonology and morphology.

Let us consider again the output sentences *Marina said that Mico left* and *Mico said that Marina left*. How can we structure their inputs so that they are distinctive, yet leave room for syntactic constraints to perform nontrivial evaluations of candidate outputs? One way to do it is to adopt the mechanism proposed in the subsection "Morpheme Order" above, namely to incorporate syntactic subtableaux into syntactic inputs. More precisely, suppose that the evaluation of the subordinate clauses may be included in the input numerations for those sentences, so that one possible input for the output *Marina said that Mico left* is {*said*, $Marina_{NOM}$, $T(that, left, Mico_{NOM})$}. Let us call such an input a **partially–structured input**. Suppose also that any change to the optimal structure of a subtableau in a partially–structured input results in a violation of a faithfulness constraint we call FAITH(SUBTAB). Then the optimal output for the partially–structured input just given is *Marina said that Mico left*, as shown in (9).

(8) FAITH(SUBTAB): The optimal structure of a subtableau is selected in the output.

(9) A tableau for *Marina said that Mico left* containing partially–structured input

said, Marina$_{\text{NOM}}$, *T*(*that*, *left*, Mico$_{\text{NOM}}$)	FAITHSUBTAB
☞ Marina said that Mico left	
Mico said that Marina left	*!

Thus, the OT perspective enables us to formulate an input which is uniquely related to a particular output (or perhaps to all synonymous outputs that make use of the same vocabulary), without trivializing the role of the constraints that evaluate syntactic structure. Of course, the output sentence *Marina said that Mico left* is still optimally (along with other outputs, such as *Mico said that Marina left*) related to the unstructured input numeration {*left*, *said*, *that*, Marina$_{\text{NOM}}$, Mico$_{\text{NOM}}$}, but that simply leads to the now–familiar situation of a given output being related to more than one input.

Given that FAITHSUBTAB is itself a constraint, the question naturally arises whether it can be violated. The answer, not surprisingly, is yes. Consider the output structure for the sentence *Marina asked when Mico left*, which is uniquely related to the partially–structured input {*asked*, *Marina*, *T*({*left*, *Q*, *when*, *Mico*})}, where *Q* is an interrogative complementizer (see Chapter 5). The highest–ranked candidate selected by the subtableau *T*({*left*, *Q*, *when*, *Mico*}) is *when did Mico leave*. However, the value of *T*({*left*, *Q*, *when*, *Mico*}) in the main numeration is the lower–ranked candidate *when Mico left*. This and similar examples show that entire subtableaux, not just substructures, appear in partially–structured inputs in syntax.

The Problem of Too Many Inputs

The arithmetic analogy given above in the subsection "Choosing the Correct Input" carries over to linguistics in another way: the "larger" the output, the larger the number of inputs that can be related to it. In syntax, and in some languages also in morphology, the size of the output is potentially unbounded. As we consider increasingly larger outputs in syntax (measured, say, by the number of phrases for which subtableaux may be appear in the input), the number of inputs that may be optimally associated with those outputs grows extremely rapidly.[11] This suggests that people are highly selective in the kinds of input–output relations they assign to sentences beyond a certain size. We conjecture that this process is also controlled by optimization, with the input "chunked" into subtableaux so that each part is the largest possible while still insuring that semantically distinct outputs are associated with distinct inputs. But this question, like many of the others raised in this volume, can only be answered with further research.

---

[11]The number of potential inputs grows exponentially with the size of the output, as measured by the number of phrases for which subtableaux may be constructed.

References

Note: Versions of certain papers in this list are available via the Rutgers Optimality Archive; see the Foreword for instructions on how to obtain them. The entries for those papers include the identification number of the paper in the archive (e.g. ROA–20 is the number for Michael Hammond's paper, Walmatjari Stress.). The papers stored in the archive may not be the latest versions, nor the specific versions cited here.

Akinlabi, Akinbiyi. 1995. Featural Affixation. In *Theoretical Approaches to African Linguistics*, ed. Akinbiyi Akinlabi, 217–237. Trenton, N.J.: Africa World Press.

Akinlabi, Akinbiyi. 1996. Featural Affixation. *Journal of Linguistics* 32:239–290.

Akmajian, Adrian, and Frank Heny. 1975. *An Introduction to the Principles of Transformational Syntax*. Cambridge, Mass.: MIT Press.

Anderson, Stephen R. 1992. *A–Morphous Morphology*. Cambridge, U.K.: Cambridge University Press.

Anderson, Stephen R. to appear. How to Put Your Clitics in Their Place, or Why the Best Account of Second–Position Phenomena May Be a Nearly Optimal One. *The Linguistic Review*. (ROA–21)

Anttila, Arto. 1995. Deriving Variation from Grammar: A Study of Finnish Genitives. Ms., Stanford University, Calif. (ROA–63)

Archangeli, Diana, and Douglas Pulleyblank. 1994. *Grounded Phonology*. Cambridge, Mass.: MIT Press.

Arnott, D.W. 1964. Downstep in the Tiv Verbal System. *African Language Studies* 5: 34–51.

Aronoff, Mark. 1976. *Word Formation in Generative Grammar*. Cambridge, Mass.: MIT Press.

Aronoff, Mark. 1993. *Morphology by Itself*. Cambridge, Mass.: MIT Press.

Bach, Emmon. 1962. The Order of Elements in a Transformational Grammar of German. *Language* 38: 263–269.

Bagemihl, Bruce. 1988. Alternate Phonologies and Morphologies. Doctoral dissertation, University of British Columbia, Vancouver.

Bagemihl, Bruce. 1991. Syllable Structure in Bella Coola. *Linguistic Inquiry* 22:589–646.

Barbosa, Pilar, Danny Fox, Paul Hagstrom, Martha McGinnis and David Pesetsky, eds. to appear. *Is the Best Good Enough?* Cambridge, Mass.: MIT Press.

Beckman, Jill N., Laura W. Dickey and Suzanne Urbanczyk, eds. 1995. *Papers in Optimality Theory: University of Massachusetts Occasional Papers* 18. Amherst, Mass.: Graduate Linguistic Student Association.

Benua, Laura. 1995. Identity Effects in Morphological Truncation. In *Papers in Optimality Theory: University of Massachusetts Occasional Papers* 18:77–136. Amherst, Mass.: Graduate Linguistic Student Association. (ROA–74)

Bierwisch, Manfred. 1963. *Grammatik des Deutschen Verbs*. Berlin: Akademie Verlag.

Bobaljik, Jonathan. 1995. Morphosyntax: The Syntax of Verbal Inflection. Doctoral dissertation, Massachusetts Institute of Technology, Cambridge.

Borowsky, Toni. 1986. Topics in the Lexical Phonology of English. Doctoral dissertation, University of Massachusetts, Amherst.

Bouchard, Denis. 1984. *On the Content of Empty Categories*. Dordrecht: Foris Publications.

Bradley, Diane, R.M. Sanchez–Casas and J.E. Garcia–Albea. 1993. Language Specific Segmentation Strategies I: Performance over Native Language Materials. *Language and Cognitive Processes* 8:197–233.

Brown, Roger, and Colin Fraser. 1963. The Acquisition of Syntax. In *Verbal Behavior and Learning: Problems and Processes*, ed. C. Cofer and B. Musgrave. New York: McGraw–Hill.

Burzio, Luigi. 1994. Weak Anaphora. In *Paths Toward Universal Grammar: Studies in Honor of Richard S. Kayne*, ed. Guglielmo Cinque, Jan Koster, Luigi Rizzi and Rafaella Zanuttini. Washington, DC: Georgetown University Press.

Bybee, Joan. 1985. *Morphology: A Study of the Relation Between Meaning and Form*. Amsterdam: John Benjamins.

Campos, Hector. 1986. Indefinite Object Drop. *Linguistic Inquiry* 17:354–359.

Carlin, Eithne. 1993. The So Language. *Afrikanistische Monographien* 2. Institut für Afrikanistik, Universität zu Köln.

Carstairs, Andrew. 1987. *Allomorphy in Inflexion*. London: Croom Helm.

Cheng, Lisa Lai Shen. 1991. On the Typology of Wh–Questions. Doctoral dissertation, Massachusetts Institute of Technology, Cambridge.

Chomsky, Noam. 1957. *Syntactic Structures*. The Hague: Mouton.

Chomsky, Noam. 1965. *Aspects of the Theory of Syntax*. Cambridge, Mass.: MIT Press.

Chomsky, Noam. 1975. *Reflections on Language*. New York: Pantheon Books.

Chomsky, Noam. 1981. *Lectures on Government and Binding*. Dordrecht: Foris Publications.

Chomsky, Noam. 1986. *Knowledge of Language: Its Nature, Origins, and Use*. New York: Praeger.

Chomsky, Noam. 1992. *A Minimalist Program for Linguistic Theory*. Massachusetts Institute of Technology Working Papers in Linguistics. Cambridge: Massachusetts Institute of Technology.

Chomsky, Noam. 1995a. Bare Phrase Structure. In *Government and Binding Theory and the Minimalist Program*, ed. Gert Webelhuth, 383–439. Oxford: Blackwell Publishers.

Chomsky, Noam. 1995b. *The Minimalist Program*. Cambridge, Mass.: MIT Press.

Chomsky, Noam, and Howard Lasnik. 1977. Filters and Control. *Linguistic Inquiry* 8:425–504.

Chomsky, Noam, and Howard Lasnik. 1993. The Theory of Principles and Parameters. In *Syntax: An International Handbook of Contemporary Research*, ed. J. Jacobs, A. von Stechow, W. Sternefeld and T. Vennemann. Berlin: Mouton de Gruyter. Reprinted in Chomsky (1995b), 13–127.

Christdas, Prathima. 1988. The Phonology and Morphology of Tamil. Doctoral dissertation, Cornell University, Ithaca, N.Y.

Cinque, Guglielmo. 1981. On the Theory of Relative Clauses and Markedness. *The Linguistic Review* 1:247–294.

Clark, Eve V. 1987. The Principle of Contrast: A Constraint on Language Acquisition. In *Mechanisms of Language Acquisition*, ed. Brian MacWhinney, 1–33. Hillsdale, N.J.: Lawrence Erlbaum.

Cole, Jennifer, and Charles W. Kisseberth. 1994. An Optimal Domains Theory of Harmony. Cognitive Science Technical Report UIUC–BI–CS–94–02, University of Illinois, Champaign–Urbana. (ROA–22)

Cole, Peter. 1982. *Imbabura Quechua*. Amsterdam: North Holland Publishing Co.

Cole, Peter. 1987. Null Objects in Universal Grammar. *Linguistic Inquiry* 18:597–612.

Comrie, Bernard, ed. 1987. *The World's Major Languages*. London: Routledge.

Crowley, Terry. 1982. *The Paamese Language of Vanuatu* (Pacific Linguistics, Series B, No. 87). Canberra: The Australian National University.

Cutler, Anne, Jacques Mehler, Dennis Norris and Juan Segui. 1983. A Language Specific Comprehension Strategy. *Nature* 304:159–160.

Cutler, Anne, Jacques Mehler, Dennis Norris and Juan Segui. 1986. The Syllable's Differing Role in the Segmentation of French and English. *Journal of Memory and Language* 25:385–400.

Cutler, Anne, and Dennis Norris. 1988. The Role of Strong Syllables in Segmentation for Lexical Access. *Journal of Experimental Psychology: Human Perception and Performance* 14:113–121.

Davis, Stuart, and Michael Hammond. 1995. Onglides in American English. *Phonology* 12:159–182.

Dell, François, and Mohammed Elmedlaoui. 1985. Syllabic Consonants and Syllabification in Imdlawn Tashlhiyt Berber. *Journal of African Languages and Linguistics* 7:105–130.

den Besten, Hans. 1983. On the Interaction of Root Transformations and Lexical Deletive Rules. In *On the Formal Syntax of the Westgermania*, ed. Werner Abraham. Amsterdam: John Benjamins.

Elzinga, Dirk. 1996. Tune–Text Association in Psalm Tunes. Ms., University of Arizona, Tucson.

Escalante, Fernando. 1990. Voice and Argument Structure in Yaqui. Doctoral dissertation, University of Arizona, Tucson.

Farrell, Patrick. 1990. Null Objects in Brazilian Portuguese. *Natural Language and Linguistic Theory* 8.3:325–346.

Fitzgerald, Colleen M. 1995. 'Poetic Meter » Morphology' in Tohono O'odham. Paper presented at the annual meeting of the Linguistic Society of America, New Orleans.

Fitzgerald, Colleen M., and Amy V. Fountain. 1995. The Optimal Account of Tohono O'odham Truncation. Ms., University of Arizona, Tucson.

Fodor, Jerry. 1983. *Modularity of Mind*. Cambridge, Mass.: MIT Press.

Fowler, Carol. 1977. Timing Control in Speech Production. Doctoral dissertation, University of Connecticut, Storrs.

Fussell, Paul. 1975. *The Great War and Modern Memory*. Oxford: Oxford University Press.

Gazdar, Gerald, Ewan Klein, Geoffrey Pullum and Ivan Sag. 1985. *Generalized Phrase Structure Grammar*. Cambridge, Mass.: Harvard University Press.

Gerken, LouAnn. 1994. Young Children's Representation of Prosodic Structure: Evidence from English–Speakers' Weak Syllable Omissions. *Journal of Memory and Language* 33:19–38.

Gnanadesikan, Amalia E. 1995. Markedness and Faithfulness Constraints in Child Phonology. Ms., Rutgers University, New Brunswick, N.J. (ROA–67)

Goldsmith, John. 1979. *Autosegmental Phonology*. New York: Garland Press.

Golston, Christopher. 1996. Direct Optimality Theory: Representation as Pure Markedness. *Language* 72:713–748.

Grimshaw, Jane. to appear. Projection, Heads and Optimality Theory. *Linguistic Inquiry*. (ROA–68)

Grimshaw, Jane, and Vieri Samek–Lodoviki. 1995. Optimal Subjects. *Papers in Optimality Theory: University of Massachusetts Occasional Papers* 18:589–606. Amherst, Mass.: Graduate Linguistic Student Association.

Guasti, Maria Teresa, and Ur Shlonsky. 1995. The Acquisition of French Relative Clauses Reconsidered. *Language Acquisition* 4:257–276.

Haegeman, Liliane. 1994. *Introduction to Government and Binding Theory*. Oxford: Blackwell Publishers.

Hammond, Michael. 1982. Foot Domain Rules and Metrical Locality. *Proceedings of the West Coast Conference on Formal Linguistics* 1:207–218.

Hammond, Michael. 1988. *Constraining Metrical Theory: A Modular Theory of Rhythm and Destressing*. New York: Garland Press.

Hammond, Michael. 1988. Templatic Transfer in Arabic Broken Plurals. *Natural Language and Linguistic Theory* 6:247–270.

Hammond, Michael. 1990. The 'Name Game' and Onset Simplification. *Phonology* 7:159–162.

Hammond, Michael. 1991. Poetic Meter and the Arboreal Grid. *Language* 67:240–259.

Hammond, Michael. 1993. Resyllabification in English. *Formal Language Studies in the Midwest* 4:104–122.

Hammond, Michael. 1994. Walmatjari Stress. Ms., University of Arizona, Tucson. (ROA–20)

Hammond, Michael. 1995. There is No Lexicon! Ms., University of Arizona, Tucson. (ROA–43)

Hammond, Michael, and Emmanuel Dupoux. to appear. Psychophonology. In *Current Trends in Phonology: Models and Methods*, ed. J. Durand and B. Laks. Manchester, U.K.: University of Salford Publications.

Harris, James W. 1983. *Syllable Structure and Stress in Spanish: A Nonlinear Analysis*. Cambridge, Mass.: MIT Press.

Hayes, Bruce. 1985. *A Metrical Theory of Stress*. New York: Garland Press.

Hayes, Bruce, and Margaret McEachern. 1996. Folk Verse Form in English. Ms., University of California, Los Angeles. (ROA–119)

Hoffmann, Carl. 1963. *A Grammar of the Margi Language*. London: Oxford University Press.

Hoonchamlong, Yuphaphann. 1984. Some Cases of Zero Pronouns in Thai: A Look from GB. Ms., University of Wisconsin, Madison.

Hoonchamlong, Yuphaphann. 1991. Some Issues in Thai Anaphora. A Government and Binding Approach. Doctoral dissertation, University of Wisconsin, Madison.

Hooper, Joan. 1972. The Syllable in Phonological Theory. *Language* 48:525–540.

Huang, C.T. James. 1984. On the Distribution and Reference of Empty Pronouns. *Linguistic Inquiry* 15:531–574.

Huang, C.T. James. 1989. Pro–Drop in Chinese. In *The Null Subject Parameter*, ed. Osvaldo Jaeggli and Kenneth Safir, 185–214. Dordrecht: Kluwer Academic Publishers.

Huang, C.T. James. 1991. Remarks on the Status of the Null Object. In *Principles and Parameters in Comparative Grammar*, ed. Robert Freiden. Cambridge, Mass.: MIT Press.

Huck, Geoffrey J., and John A. Goldsmith. 1995. *Ideology and Linguistic Theory: Noam Chomsky and the Deep Structure Debates*. London and New York: Routledge.

Hyman, Larry M. 1975. *Phonology: Theory and Analysis*. New York: Holt, Rinehart and Winston.

Itô, Junko. 1989. A Prosodic Theory of Epenthesis. *Natural Language and Linguistic Theory* 7:217–260.

Itô, Junko, Rolf Armin Mester and Jaye Padgett. 1995. Licensing and Underspecification in Optimality Theory. *Linguistic Inquiry* 26:571–613. (ROA–38)

Jaeggli, Osvaldo, and Kenneth Safir, eds. 1989. *The Null Subject Parameter*. Dordrecht: Kluwer Academic Publishers.

Jiang–King, Ping. 1996. Tone–Vowel Interaction in Optimality Theory. Doctoral dissertation, University of British Columbia, Vancouver.

Kahn, Daniel. 1976. Syllable–Based Generalizations in English Phonology. Doctoral dissertation, Massachusetts Institute of Technology, Cambridge, Mass.

Kayne, Richard S. 1977. French Relative *Que*. In *Current Studies in Romance Linguistics*, ed. Marta Luján and Fritz Hensey. Washington D.C.: Georgetown University Press.

Keer, Ed and Eric Bakovic. 1996. Sometimes Two Heads Are Better than One (or None). Talk given at Rutgers University–University of Massachusetts Joint Class Meeting II, Amherst, Mass.

Kenstowicz, Michael. 1994. *Phonology in Generative Grammar*. Cambridge: Blackwell Publishers.

Kenstowicz, Michael, and Charles Kisseberth. 1979. *Generative Phonology: Description and Theory*. New York: Academic Press.

Keyser, Samuel Jay. 1975. A Partial History of the Relative Clause in English. In *Papers in the History and Structure of English*, ed. Jane Grimshaw, 1–43. Amherst: University of Massachusetts.

Kiparsky Paul. 1977. The Rhythmic Structure of English Verse. *Linguistic Inquiry* 8:189–247.

Kiparsky, Paul. 1982. From Cyclic to Lexical Phonology. *The Structure of Phonological Representations*, ed. Harry van der Hulst and Neil Smith, 131–175. Dordrecht, NL: Foris.

Kiparsky, Paul. 1985. Some Consequences of Lexical Phonology. *Phonology Yearbook* 2:85–138.

Koizumi, Masatoshi. 1995. Phrase Structure in Minimalist Syntax. Doctoral dissertation, Massachusetts Institute of Technology, Cambridge.

Kuroda, Sige–Yuki. 1983. What Can Japanese Say about Government and Binding? *Proceedings of the West Coast Conference on Formal Linguistics* 2:153–164.

Labelle, Marie. 1990. Predication, WH–Movement, and the Development of Relative Clauses. *Language Acquisition* 1:95–120.

Ladefoged, Peter. 1968. *A Phonetic Study of West African Languages*. Cambridge, U.K.: Cambridge University Press.

Ladefoged, Peter. 1975. *A Course in Phonetics*. New York: Harcourt Brace Jovanovich.

Ladefoged, Peter, and Ian Maddieson. 1996. *The Sounds of the World's Languages*. Oxford: Blackwell Publishers.

Legendre, Geraldine, Colin Wilson, Paul Smolensky, Kristen Homer and William Raymond. to appear. Optimality and *Wh*–Extraction. In Barbosa et al. (to appear). (ROA–85)

Lehiste, Ilse. 1970. *Suprasegmentals*. Cambridge, Mass.: MIT Press.

Maddieson, Ian. 1983. The Analysis of Complex Phonetic Elements in Bura and the Syllable. *Studies in African Linguistics* 14:285–310.

Manzini, Rita. 1983. On Control and Control Theory. *Linguistic Inquiry* 14:421–446.

Massar, Andrea. 1996. Syllable Omission and the Prosodic Structure of Two–Year–Olds. Ms., University of Arizona, Tucson.

Matthews, Peter H. 1972a. *Inflectional Morphology: A Theoretical Study Based on Aspects of Latin Verb Conjugation*. Cambridge, U.K.: Cambridge University Press.

Matthews, Peter H. 1972b. Huave Verb Morphology: Some Comments from a Non–Tagmemic Perspective. *International Journal of American Linguistics* 38:96–118.

McCarthy John J. 1982. Prosodic Structure and Expletive Infixation. *Language* 58:574–590.

McCarthy, John J. 1983. Consonantal Morphology in the Chaha Verb. *Proceedings of West Coast Conference on Formal Linguistics* 2:176–188.

McCarthy, John J. 1989. Linear Order in Phonological Representation. *Linguistic Inquiry* 20:71–99.

McCarthy, John J. 1995. Extensions of Faithfulness: Rotuman Revisited. Ms., University of Massachusetts, Amherst. (ROA–110)

McCarthy, John J., and Alan S. Prince. 1990. Foot and Word in Prosodic Morphology: the Arabic Broken Plural. *Natural Language and Linguistic Theory* 8:209–283.

McCarthy, John J., and Alan S. Prince. 1993a. Prosodic Morphology I: Constraint Interaction and Satisfaction. Ms., University of Massachusetts, Amherst, and Rutgers University, New Brunswick, N.J.

McCarthy, John J., and Alan S. Prince. 1993b. Generalized Alignment. In *Yearbook of Morphology 1993*, 79–154. (ROA–7)

McCarthy, John J., and Alan S. Prince. 1994. The Emergence of the Unmarked: Optimality in Prosodic Morphology. *Proceedings of the North East Linguistic Society* 24:333–379. (ROA–13)

McCarthy, John J., and Alan S. Prince. 1995. Faithfulness and Reduplicative Identity. In *Papers in Optimality Theory: University of Massachusetts Occasional Papers* 18:249–384. Amherst Mass.: Graduate Linguistic Student Association. (ROA–60)

McCarthy, John J., and Alan S. Prince. to appear. Reduplicative Identity. *Proceedings of the 1996 Utrecht Prosodic Morphology Workshop*. Utrecht, The Netherlands.

McCloskey, James. 1995. *Wh*–Movement and Quantifier Float in an Irish English. Paper presented at 26th North East Linguistics Society meeting, Massachusetts Institute of Technology, Cambridge.

Meador, Diane. 1996. The Minimal Word Hypothesis: A Speech Segmentation Strategy. Doctoral dissertation, University of Arizona, Tucson.

Mohanan, K.P. 1993. Fields of Attraction in Phonology. In *The Last Phonological Rule*, ed. John Goldsmith, 61–116. Chicago, Ill.: University of Chicago Press.

Montalbetti, Mario. 1984. After Binding. Doctoral dissertation, Massachusetts Institute of Technology, Cambridge, Mass.

Myers, Scott. 1987. Vowel Shortening in English. *Natural Language and Linguistic Theory* 5:485–518.

Myers, Scott. 1990. *Tone and the Structure of Words in Shona*. New York: Garland Press.

Myers, Scott. 1995. OCP Effects in Optimality Theory. Ms., University of Texas, Austin. (ROA–6)

Newman, Stanley. 1944. *The Yokuts Language* (Viking Fund Publications in Anthropology 2). New York: The Viking Fund.

Ohala, Diane. 1996. Cluster Reduction and Constraints in Acquisition. Doctoral dissertation, University of Arizona, Tucson.

Padgett, Jaye. 1994. Stricture and Nasal Place Assimilation. *Natural Language and Linguistic Theory* 12:465–513.

Padgett, Jaye. to appear. Partial Class Behavior and Nasal Place Assimilation. *Proceedings of the Arizona Phonology Conference: Workshop on Features in Optimality Theory*. University of Arizona, Tucson. (ROA–113)

Pater, Joe. to appear. Austronesian Nasal Substitution and Other NC Effects. *Proceedings of the 1996 Utrecht Prosodic Morphology Workshop*. Utrecht, The Netherlands. (ROA–92)

Pérez, Patricia. 1992. Gradient Sonority and Harmonic Foot Repair in English Syncope. *Coyote Papers* 8:118–142. University of Arizona, Tucson.

Pesetsky, David. to appear. Some Optimality Principles of Sentence Pronunciation. In Barbosa et al. (to appear). (ROA–42)

Peters, Ann M., and Lise Menn. 1993. False Starts and Filler Syllables: Ways to Learn Grammatical Morphemes. *Language* 69:742–777.

Pinker, Steven. 1984. *Language Learnability and Language Development*. Cambridge, Mass.: Harvard University Press.

Pinker, Steven. 1994. *The Language Instinct*. New York: W. Morrow and Company.

Pinker, Steven. 1995. Why the Child Holded the Baby Rabbits: A Case Study in Language Acquisition. In *An Invitation to Cognitive Science, Volume 1: Language*, 2nd edition, ed. Lila Gleitman and Mark Liberman, 107–137. Cambridge, MA: MIT Press.

Plank, Frans, ed. 1991. *Paradigms: The Economy of Inflection*. Berlin: Mouton de Gruyter.

Pollard, Carl, and Ivan A. Sag. 1992. Anaphors in English and the Scope of Binding Theory. *Linguistic Inquiry* 23.2:261–303.

Pollard, Carl, and Ivan A. Sag. 1994. *Head–Driven Phrase Structure Grammar*. Chicago, Ill.: University of Chicago Press.

Prince, Alan S., and Paul Smolensky. 1991. Optimality. Talk given at Arizona Phonology Conference 3, University of Arizona, Tucson.

Prince, Alan S., and Paul Smolensky. 1993. *Optimality Theory: Constraint Interaction in Generative Grammar*, RuCCs Technical Report #2, Rutgers University Center for Cognitive Science, Piscataway, N.J. [to appear, Cambridge, Mass.: MIT Press].

Pukui, Mary Kawena, and Samuel H. Elbert. 1957. *Hawaiian Dictionary*. Honolulu: University Press of Hawaii.

Pulleyblank, Douglas. 1986. *Tone in Lexical Phonology*. Dordrecht: Reidel.

Pulleyblank, Douglas. 1993. Vowel Harmony and Optimality Theory. *Proceedings of the Workshop on Phonology*, 1–18. University of Coimbra, Portugal, Associação Portuguesa de Linguística.

Pulleyblank, Douglas. 1994. Neutral Vowels in Optimality Theory: A Comparison of Yoruba and Wolof. Ms., University of British Columbia, Vancouver.

Pulleyblank, Douglas. 1995. Feature Geometry and Underspecification. In *Frontiers of Phonology*, ed. Jacques Durand and Francis Katamba, 3–33. London: Longman.

Radford, Andrew. 1988. *Transformational Grammar: A First Course*. Cambridge, U.K.: Cambridge University Press.

Radford, Andrew. 1990. *Syntactic Theory and the Acquisition of English Syntax*. Oxford: Blackwell Publishers.

Reinhart, Tanya, and Eric Reuland. 1993. Reflexivity. *Linguistic Inquiry* 24:657–720.

Rizzi, Luigi. 1986. Null Objects in Italian and the Theory of pro. *Linguistic Inquiry* 17:501–557.

Russell, Kevin. 1995. Morphemes and Candidates in Optimality Theory. Ms., University of Manitoba, Winnipeg. (ROA–44)

Sagey, Elizabeth. 1986. The Representation of Features and Relations in Non–Linear Phonology. Doctoral dissertation, Massachusetts Institute of Technology, Cambridge, Mass.

Selkirk, Elizabeth. 1982. The Syllable. In *The Structure of Phonological Representations*, vol. II, ed. Harry van der Hulst and Norval Smith, 337–383. Dordrecht: Foris.

Sells, Peter. 1985. *Lectures on Contemporary Syntactic Theories*. Stanford, Calif.: Center for the Study of Language and Information.

Speas, Margaret. 1995. Generalized Control and Null Objects in Optimality Theory. In *Papers in Optimality Theory: University of Massachusetts Occasional Papers* 18. Amherst, Mass.: Graduate Linguistic Student Association.

Stairs, Emily F., and Barbara Erickson. 1969. Huave Verb Morphology. *International Journal of American Linguistics* 35:38–53.

Stemberger, Joseph. 1981. Morphological Haplology. *Language* 67:791–817.

Tesar, Bruce. 1995. Computational Optimality Theory. Doctoral dissertation, University of Colorado, Boulder. (ROA–90)

Treiman, Rebecca, and C. Danis. 1988. Syllabification of Intervocalic Consonants. *Journal of Memory and Language* 27:87–104.

Treiman, Rebecca, and A. Zukowski. 1990. Toward an Understanding of English Syllabification. *Journal of Memory and Language* 29:66–85.

Ward, Ida C. 1952. *An Introduction to the Yoruba Language*. Cambridge, U.K.: W. Heffer and Sons.

Welmers, William E. 1970. Igbo Tonology. *Studies in African Linguistics* 1:255–278.

REFERENCES

Wonderly, William. 1951. Zoque I, II, III, IV. *International Journal of American Linguistics* 17:1–9, 105–123, 137–162, 235–251.

Wonderly, William. 1952. Zoque V, VI. *International Journal of American Linguistics* 18:35–48, 189–202.

Xu, Liejiong. 1986. Free Empty Category. *Linguistic Inquiry* 17:75–93.

Xu, Liejiong, and D. Terence Langendoen. 1985. Topic Structures in Chinese. *Language* 61.1–27.

Yip, Moira. 1995. Identity Avoidance in Phonology and Morphology. Ms., University of California, Irvine. (ROA–82)

Yoon, J. 1985. On the Treatment of Empty Categories in Topic Prominent Languages. Ms., University of Illinois, Champaign–Urbana.

Zwicky, Arnold. 1972. Note on a Phonological Hierarchy in English. In *Linguistic Change and Generative Theory: Essays from the UCLA Conference on Historical Linguistics in the Perspective of Transformational Theory*, ed. Robert Stockwell and Ronald Macaulay, 275–301. Bloomington: Indiana University Press.

Index